The Snow Booklet

A Guide to the Science, Climatology, and Measurement of Snow in the United States

Nolan J. Doesken and Arthur Judson

Colorado Climate Center
Department of Atmospheric Science
Colorado State University
Fort Collins, CO 80523

With support from:

• U.S. Department of Interior Bureau of Reclamation, Flood Hydrology

National Climatic Data

Conservation Service,

ation

State University

Front Cover: Winter sun filters through a snow-covered Colorado aspen grove.
Photo by Daniel B. Glanz

About the Authors

Nolan Doesken

Nolan Doesken's love for snow developed at an early age. As a child, he could not sleep at night if snow was falling or predicted. Much of Doesken's appreciation for snow came from his years as a paperboy in a small town in central Illinois. "I particularly remember a Sunday morning in March or April when I awoke to an unexpected four inches of snow. The temperature was close to 32°F, so I figured the snow would be heavy and wet. But, to my amazement, it was like goose down. As I loaded my papers, a car drove past the house at a low speed. A cloud of fluff lifted into the air and when the car was gone, the street had been blown clear. That's when I learned that the ten to one 'rule' didn't always apply." Since 1977, Doesken has been the Assistant State Climatologist at the Colorado Climate Center at Colorado State University. He also is a National Weather Service cooperative weather observer. Doesken closely works with National Weather Service personnel to improve the quality of weather data for climatic applications.

Arthur Judson

Snowbursts that frequented his father's logging camps on Tug Hill, New York, fascinated Art Judson. As a young Marine, he chased snowstorms in California's San Bernadino mountains, then found deeper snows in the Rockies and Cascades. After obtaining a B.S. in Forestry from Oregon State University in 1960, he started chasing avalanches as a Forest Service snow ranger in Colorado. He later became a career avalanche forecaster and snow scientist with the Forest Service Avalanche Project. "Jud," as he is known to his friends, organized a special network to provide weather, snow, and avalanche data for avalanche forecasting and warning across the mountainous West. He founded Colorado's Avalanche Warning Program, developed an avalanche forecasting model, and worked to establish a warning service in Alaska. Currently, Jud lives in Steamboat Springs, Colorado (average annual snowfall 171 inches with 300-600 inches falling on the nearby mountains), where he continues to enjoy snow greatly.

Snow – water in the form of countless ice crystals – has many intriguing traits. As this photo clearly demonstrates, snow is a viscoelastic material able to deform without breaking at temperatures near its melting point.

Photo by Milly Judson

Table of Contents

Folklore about Snow

"As high as the summer weeds did grow, that is how deep shall bank the snow."

According to Kentucky folklore, the number of fogs observed during the month of August is the number of snows that fall during the following winter.

Most early American folklore forecasts predicted early or hard winters. While winter snow varies greatly from year to year, anticipating an early and a hard winter was very much in the best interest of pioneers and early settlers whose very survival depended on how well prepared they were for the winter season.

Photo by Jack G. Jones, Natural Resources Conservation Service

Preface

Snow, snow, snow – to some it is a nasty four-letter word. To others, it is the very essence of the mysteries of nature. Snow stimulates curiosity among scientists. It brings people together who might otherwise never meet. To the weather observer, it is a source of seasonal challenge. So much is known about snow, yet it comes as a wondrous surprise to each new generation.

This book is written for climatological observers and their managers, snow-fighters, urban-planners, winter-recreationists, and all who find in snow a sense of inspiration and awe. Our goal is to better inform climatological observers and others about snow and its characteristics. Information and instructions for consistent measurements of the phenomena are included so that more and better data about snow are available in the future. While there is no substitute for experience with snow, knowledge of the behavior of ice crystals and snow grains will help those whose lifestyle, work, or hobby is affected by snow.

Calvin and Hobbes   by Bill Watterson

Photo credits: Bob Winsett, Keystone (top).
Grant Goodge (lower left). Beaver Creek,
Colorado Ski Country USA (lower center).
Charles Kuster (lower right).

The Power and Beauty of Snow

Snow always has challenged humans, but also has fascinated and delighted us. It is so beautiful, so innocent in its whiteness, and so appealing. It transforms the landscape around us, even changing the sounds we hear and the odors we smell. For some, anticipation of an approaching snowstorm is like waiting for Santa Claus on Christmas Eve. But the beauty and joy always is balanced by apprehension and hazard. Today, as in the past, basic tasks of life such as securing food, staying warm and dry, and performing normal work become more difficult when deep snow is present or falling. Transportation, whether by foot, horse, train, car, or plane, is especially vulnerable.

Snow is secretly dynamic. Its physical properties constantly change in time and space. Like the Will O' the Wisp, it may appear almost magically, modify its form, and then quickly disappear. Both falling snow and snow on the ground are subject to change. What begins as a fluffy blanket of soft snow may become dense and mushy in less than a day. By the next day, it may be solid and crusty and become crumbly and granular several days later.

Snow is an inestimable resource that impacts the national economy. It provides life-giving water. In much of the mountain West, between 60 and 75 percent of annual precipitation falls in the form of snow. One-fifth of annual precipitation in the Northern Great Plains falls as snow. In that region snow diminishes soil freezing, guards crops from winterkill, and recharges soil moisture critical to grain production. The snow cover has biological significance as home to many small animals and plants. In the far north, snow even is used for shelter by indigenous peoples. Snow also is the key ingredient for the burgeoning winter sports industry. Its depth on ski runs is so critical that people found a way to make more of it when not enough falls.

Photo by Ken Dewey

Human response and adaptation to snow are remarkable. A growing majority of the U.S. population now lives in areas that receive very little snowfall. Yet each year, millions of Americans travel to some of the snowiest parts of the U.S. and spend hundreds-of-millions of dollars to play in the snow, to ski, snowmobile, and take moonlit sleigh rides. Hundreds-of-thousands of others migrate south away from the snowy Great Lakes, the Upper Midwest, and New England, having had their fill of snow shovels and snarled traffic.

Snow affects global climate. The ability of snow to reflect vast amounts of incoming solar radiation helps drive global wind patterns. At high latitudes snow's radiative properties chill large airmasses that slip southward into temperate zones causing storms and precipitation. The ability of snow to reflect solar radiation and to rapidly cool air sharpens winter's cold and reinforces temperature inversions that trap harmful pollutants near the ground.

Snow's adverse effects are legend. Snow collapses roofs and damages trees. Melting snows soak fields, flood valleys, delay planting, and cut crop yields. Snowstorms block roads and isolate towns and cities. Vital services are curtailed. Surface and air traffic grind to a halt. Whole cities become paralyzed. Business stops. People become trapped — some die. Some say it can't snow enough, but when it does snow the price may be high.

A wagon train with eighty-one settlers traveling from Illinois to the west coast became trapped in a blizzard near Donner Lake in the Sierras in 1846. Rescuers reached them four months later. Forty-six settlers died.

An epic blizzard in 1888 killed 400 people, half of them in New York City. This storm wreaked havoc from Chesapeake Bay to Maine.

A great snowstorm in Washington State's Cascade Mountains in 1910 (11 feet of snow in nine days) started simultaneous avalanches that swept seven locomotives and two trains down a mountain canyon at Wellington, Washington. Ninety-six people died.

A 28-inch snowfall in Washington D.C. in 1922 collapsed a movie-theater roof, killing 100.

In 1941, a blizzard killed 71 people in eastern North Dakota and Minnesota.

A 1967 snowstorm struck Chicago and caused the deaths of 45 people and estimated economic losses to area businesses of $150 million.

Floodwater from snowmelt in the Upper Midwest in 1969 killed ten and caused around $100 million in damages. Barge transportation also was crippled for weeks.

A 1978 snowstorm in New England killed 60 people, with economic losses exceeding $1 billion.

A March snowstorm in 1993 paralyzed a huge area from Alabama to Georgia northward through New England. Every airport from Atlanta to Maine was closed during the storm.

In the four winters ending in April 1995, snow avalanches in the United States killed 89 people and injured 72 others.

$$$

Christmas Day 1982 in Denver, Colorado, after a record-breaking Christmas Eve blizzard deposited two feet of snow.

© 1984 James Wiesmueller; used with permission

There are huge routine annual costs associated with snow. The cost of snow removal for streets and highways across the U.S. easily exceeds $2 billion annually. This doesn't include the millions of dollars privately spent clearing parking lots, driveways, and walks to keep commerce open for business. Eight-million-tons of salt at a cost of about $250 million (1996 estimate) are spread on our streets. Despite huge efforts to clear snow and ice, tens-of-thousands of traffic accidents and personal slip-and-fall incidents still occur that claim lives and generate hundreds-of-millions of dollars in medical costs.

Whatever the benefits and costs, there is a compelling need for better snow data for a variety of users. We hope the information in this booklet will help climate observers and users of climatological information to better understand snow and its characteristics.

The Science of Snow

What Is Snow?

Imagine, for a moment, walking through a forest on a cloudy, late-autumn afternoon. In total silence the first snow of the season begins to fall. Temperatures are slightly above freezing. Snowflakes two inches across float and spin slowly to the ground and melt. You stand watching, wondering, mesmerized – unconcerned that each of these flakes (polycrystals) are composed of tens, perhaps hundreds, of individual snow crystals. You see the astonishing intricacy of the clusters. Curious, you hold out your arm. There on your sleeve you observe symmetric, hexagonal (six-sided) branched fern-like crystals – the classic dendrite – you also notice stars and plates. You spot a twelve-pointed star made up of two hexagonal stellar crystals pressed together. A few crystals appear malformed, like mutant genes with tiny imperfections. In a blur, a heavily rimed crystal falls through the others and bounces lightly on the ground – graupel.

Photo by Grant Goodge

Magnified view of a single snow crystal.
Photo by Richard L. Armstrong, University of Colorado

Later, as the ground turns white, the snowflakes seem to change. Needle and column-shaped crystals now land on your sleeve. You are amazed by the variety. What a collage! What could possibly make it snow like this?

The origin of snow lies in subfreezing clouds where water exists in all three states of matter: liquid, vapor, and solid. Moisture for such clouds comes from oceans, large lakes, soil, and plants. After evaporation and transpiration occur from these sources, the water is transported as vapor through the atmosphere. The amount flowing across land is large; it has been estimated to be six times more than the water carrying capacity of all continental U.S. rivers combined. Only a fraction of this is available for precipitation and less than that for making snow.

A subtle characteristic of water vapor is shown in Figure 1 and is essential to the process of converting invisible water vapor into visible clouds and snow. There is a limit to how much water vapor can be added to the atmosphere, which is related to the temperature and pressure. As air rises, it expands and cools. Its capacity for water vapor decreases with lower temperatures, so as air cools, saturation eventually occurs. One-hundred percent relative humidity is achieved when the air has reached its capacity for water vapor at that temperature. Near saturation, tiny salt particles and droplets of sulfuric and nitric acid act as condensation nuclei. Water vapor is deposited on these tiny nuclei to form cloud droplets. When the air cools to below the freezing point, most droplets remain as liquid, even at temperatures well below zero (Fahrenheit). The term for such droplets is "supercooled." Simultaneously, as long as the air remains saturated, some water vapor changes directly to the solid. In this way, ice condenses onto tiny clay particles and kaolinites dispersed by wind erosion. These particles, known as freezing nuclei, are an essential part of the precipitation process. Continued cooling or continued addition of moisture produces a condition known as supersaturation and is necessary for maintaining the snow-formation process.

The cloud now contains vapor, tiny ice crystals, and cloud droplets. The crystals grow at the expense of the water droplets because the vapor pressure over supercooled water is somewhat greater than the vapor pressure over ice at the same temperature (note inset on Figure 1). This process is difficult to explain simply, but is a very important part of cloud physics and precipitation formation. A transfer of water molecules from droplets to vapor to ice crystals evolves. Snow is born.

The type of crystals that form depend on humidity and temperature. Temperature mainly determines the type of crystal while the degree of supersaturation plays a secondary role. With the exception of warm temperature crystals – needles, cups, and graupel – crystal types occur over a broad temperature range. The same cloud can produce many different types of crystals at the same time, but most will have hexagonal symmetry (in one plane) because of the molecular arrangement of the atoms in the crystal lattice. Crystal structure varies from simple, elementary forms to astonishingly complex and intricate lattice networks.

figure 1

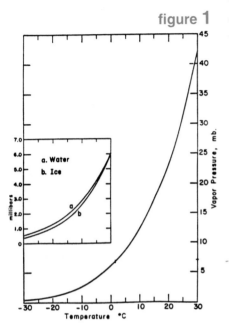

Saturation vapor pressure increases with temperature. At temperatures below the freezing point of water, the vapor pressure is greater over liquid water than over ice, key to the snow-formation process.

figure 2

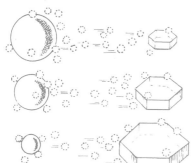

Water droplet   Ice crystal

High vapor pressure   Low vapor pressure

Vapor moves from droplets to ice crystals because the vapor pressure over water is higher than the vapor pressure over ice at the same temperature.

A localized snow shower over northern Colorado in March 1993.

Photo by Don Teem, NWS Cooperative Weather Observer at Rand, Colorado

With further growth, crystals attain sufficient size and mass to begin falling earthward. Depending on the size and shape, fall velocities of individual crystals vary from one to about six-feet per second. Some float, while others appear to streak downward. Some collide with the super-cooled droplets that instantly freeze on impact, imparting further mass and a grey cast to crystals. These accretions are called rime. Rimed crystals fall more quickly to the ground than unrimed crystals and are associated with denser snowfalls. The fastest falling crystals, those heavily rimed, will appear to the observer's fixed gaze as white streaks or snow flakes with "tails."

Many snow crystals melt or evaporate before they ever reach the ground. Those that reach the ground may have a complex history of growth and decay spanning a time period from a few minutes to as much as several hours. When they eventually reach the ground they may appear quite different from textbook examples. To survive, the crystals must spend most of their lives in a saturated to supersaturated environment, with respect to ice. This requires moisture, subfreezing temperatures, and a means for cooling the air.

The primary means for cooling air to saturation is to lift it. There are several lifting processes that lead to snow formation. These processes may operate independently or in combination.

1) Orographic lifting: air is forced upward over elevated terrain. This is a primary lifting mechanism in the western United States and explains why snowfall patterns generally follow elevation contours. Even slightly elevated topography can enhance snowfall.

figure 3

Type of Particle	Graphic Symbol	Examples	
Plate			
Stellar crystal			
Column			
Needle			
Spatial dendrite			
Capped column			
Irregular crystal			
Graupel			
Ice pellet			
Hail			

Common snow crystal forms from the international classification for solid precipitation.

figure 4

Ice crystal type and shape are determined primarily by the temperature within the cloud, but other factors such as the amount of supersaturation (excess water vapor) also make a difference as is shown in this classic diagram from SNOW CRYSTALS by Ukichiro Nakaya[7]. Copyright © 1954 by the President and Fellows of Harvard College. Reprinted by permission of Harvard University Press.

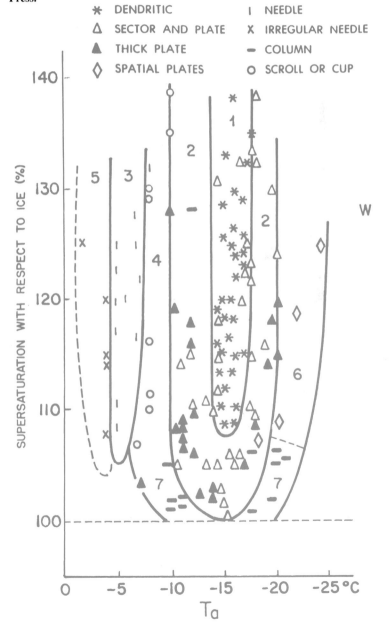

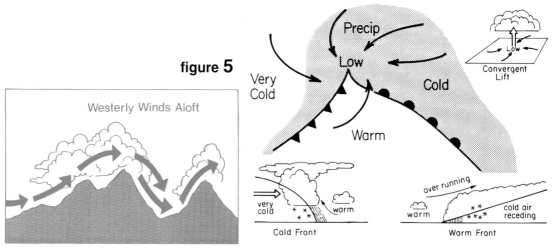

figure 5

Westerly Winds Aloft

Schematic of mechanisms that lift and cool air in the atmosphere to produce clouds and precipitation.

2) Frontal: warm air rides over or is lifted by colder air. Across the eastern half of the United States, this is a major mechanism for producing snow, as well as other forms of winter precipitation, when warm, moist air from the Gulf of Mexico slides up and over shallow layers of cold air. Vertical motions are gradual, but long-lasting and widespread and accompanied by plentiful moisture. Cold fronts may produce more rapid lifting and cause intense snow showers or squalls with graupel and other rimed crystals, but the snow is typically brief and not associated with large snowstorms.

3) Convergence: air flows inward toward low pressure with excess flows forced upward. This is a very important snowmaker for the country as a whole. There are preferred areas in the United States for low pressure areas to form and track (Figure 18), but all areas of the country are affected.

figure 6

A view from the GOES-1 geostationary satellite on February 23, 1977 at 1900 UTC provides a spectacular example of the various lifting processes at work in the atmosphere. Clouds associated with a cold front, warm front, and cyclonic convergence are evident across the central U.S. while the West experiences orographic precipitation in several mountainous areas.

Photo by NOAA/NESDIS/NCDC Asheville, North Carolina

4) Convection: sometimes the atmosphere gets cooler with height at such a rate that if air is forced to rise for any reason, that rising air will be warmer and less dense than the air around it. This results in buoyancy that will cause the air to continue to rise quite rapidly until it reaches a level where it no longer is warmer than the air around it. Convective motions produce local cumulus-type clouds and locally very heavy precipitation. Heavy snows near warm bodies of water often are enhanced in this way.

5) Other: the complex three-dimensional motions in the atmosphere along with vertical temperature gradients, may produce or enhance upward motions in a variety of ways. Visit a weather forecast office during the winter and you will hear forecasters use strange jargon like "jet streaks" and "cold air damming" as they try to determine where and when snow will fall.

Snow develops in advance of an approaching wintertime warm front.

Photos by Grant Goodge

It also is possible to achieve saturation by adding moisture to the air with little or no vertical lifting. This takes place when water from lakes and oceans evaporates into a cold air mass above. Lake-effect snows are enhanced in this way. Cooling towers and cooling lakes from large power-generation plants also are known to create highly localized snowstorms by this means.

Density Comparison

Water

1

Very Dense
Snow

0.40

Ice

0.917

Fresh Fallen
Snow

.07 **figure 7**

Relationship between the
density of water, solid ice, and
snow. The shaded area within
each cube shows the volume
that would be occupied by
liquid water, if the contents of
the cube melted.

Characteristics of Fresh Snow

Crystals are subject to constant change. Snow crystals falling through above-freezing layers begin to melt. This causes loss of edges, facets, and vertices. Turbulent winds near the ground cause shattering collisions that drastically alter crystal structure. Some crystals are dashed on contact with trees, rocks, and other objects. Others break on contact with the snow surface. Wind moves snow in three modes: rolling along the snow surface, bouncing (saltation) from the surface into the air and back onto the snow again, and in the air (turbulent suspension). About 90 percent of the total wind-induced snow in motion takes place in the first two feet above the ground[8], and although light snow particles may be lifted to heights over 300 feet, such spectacular displays contribute little to the total amount drifted. In blowing snow events, tiny snow particles pass right through key holes in doors, depositing enough snow inside to temporarily block entry.

New deposits of snow on the ground are what we commonly call snowfall. Although the accumulated depth of snowfall is measured and frequently mentioned, snow has many other intriguing attributes.

One of the most important and useful properties of fresh snow is its density – the relationship between snow depth and its water equivalent. It is an important parameter in avalanche forecasting, snowmelt flood forecasting, and snow management. For weather observers, climatologists, and data users, snow density helps characterize what the snow was like and also gives an index for evaluating data quality. The weather observer who understands new snow density and knows the range of densities that commonly occur will provide better data than would someone without this knowledge.

Density is mass per unit volume and is sometimes expressed in grams per cubic centimeter. Because density in those units is numerically equal to the water-to-snow depth ratio (water content in inches divided by snow depth in inches), this dimensionless number, technically called specific gravity, is used for density throughout this booklet for convenience. Most climate observers do not have the snow density kits used by professional snow workers, but they can determine the density of their own snow samples by using the ratio.

The density of water is one and pure ice is 0.917. The upper density limit for new snow is near 0.40, that is equivalent to four inches of water from ten inches of snow, while the lower limit is near 0.01 (one inch of water from one hundred inches of snow). "Ten-to-one" snow (ten inches of snow with a water content of one inch) has a density of 0.10.

Figure eight shows that regardless of location, latitude, and elevation, most new snow densities fall in the range from 0.04 to 0.10, with peak frequencies centered between 0.06 and 0.09. With many years of data, such as shown in figure eight, the frequency and limits of densities above 0.10 become apparent; the shape of the "tail" of the distributions is a distinct feature of each snow climate. In this "tail," one finds that densities above 0.10

figure 8

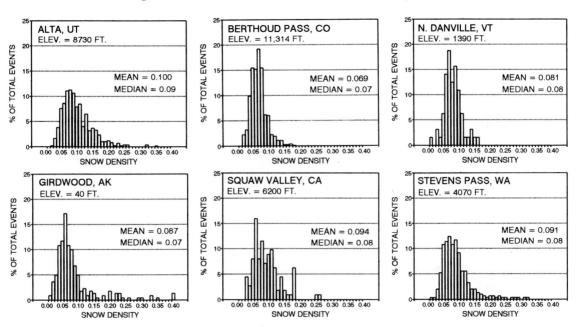

The distribution of snow density of 24-hour new snowfall totals at various U.S. locations. Alta, Girdwood, Squaw Valley and Stevens Pass after LaChapelle (1962)[6]. Water contents were determined from snowboard core samples unaffected by melt. Incidents of sleet and freezing rain were not included.

and even 0.20 are common at places like Alta, Utah; Stevens Pass, Washington; and Girdwood, Alaska. Many of these higher density values occur in snow that is dry. Graupel and needle crystals in particular produce high densities. The data also demonstrate the futility of trying to define snow density with one number such as 0.10, because such a wide range of densities is expected. Not shown, is the fact that individual daily new snow densities vary widely from one day to the next, and on the shortest time scale, from one moment to another. It is known, however, that average monthly densities (core data) commonly range from about 0.06 to 0.11. Average monthly values below about 0.05 very likely are the product of measurement error.

Anyone who has shoveled driveways and sidewalks or driven snowplows has some appreciation for snow density. Your car may successfully push through a foot or more of low-density snow (0.06 or less) but immediately gets stuck in lesser amounts of dense snow (0.10 or greater). It is well worth the effort to learn about snow density. Use your water equivalents and new snow depths to discover more of this important snow property.

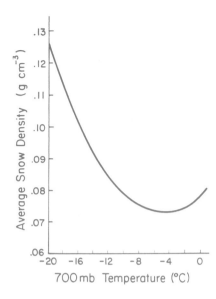

Daily densities below about 0.04, the domain of "wild snow," are rare, but exciting. Densities as low as 0.01 have been observed. Wild snow is so light it rises rapidly at the stomp of a foot or ski. It is possible to walk with ease through large quantities of it.

Factors affecting new snow density are atmospheric conditions during crystal formation and descent, and conditions experienced while landing on the surface. Warm temperatures, high winds, heavily rimed and/or small crystals all favor high density. Low density snows usually require very light winds. They occur with unrimed dendritic or plate crystals with lacy structures. Snow densities generally decrease as temperatures get colder. However, below about 10°F densities increase again and may reach 0.12 or greater since crystals at very cold temperatures are commonly small and pack into high density layers as they accumulate on the ground. This characteristic of fresh snow comes as a great surprise to many who have little experience with cold-temperature snows.

Seasonally, highest new-snow densities occur in spring and fall when snow temperatures are near the melting point, but even then large variations are common. Warm temperatures and increased solar radiation during this part of the year contribute to elevated densities.

Fresh snow has many other interesting and important properties. Fresh, dry snow can reflect more than 90 percent of the sunlight that strikes the surface (no wonder we need sun glasses). By comparison, wet snow may reflect less than 60 percent. Conversely, snow readily absorbs and emits terrestrial (long-wave) radiation. Under cloudy conditions, it quickly establishes thermal equilibrium with the air above it, but under clear skies snow surface temperatures are considerably colder than the air a few feet above it. Snow is a good insulator. A thick layer of fresh snow on your roof can save you money on your heating bill, especially if you didn't already have much attic insulation. There is also the matter of making snowballs. The key element for packable snow is that it be near the freezing point. Interestingly, fresh snow does not have to contain liquid water in order to pack into a ball. At temperatures much below freezing, pressures needed to pack snow are so great that the grain-to-grain contacts fracture and the material crumbles.

Snow on the Ground

The accumulation of old and new snow on the ground is known as snowpack. The term implies continuous or nearly continuous snow cover composed of more than one layer, but it can be used to describe any snow covering the ground. Snowpacks develop and dissolve in undulating patterns. Quietly, they spread from north to south and from high to low elevation as

Photo by Daniel B. Glanz

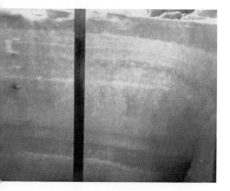

A pit wall dug in the snowpack exposes the history of snow accumulation and reveals ongoing changes in crystal structure.

Photo by Richard L. Armstrong, University of Colorado

winter progresses. By November, a discontinuous patchwork of snow is often observed, dictated by climate, storm tracks, and local effects. Looking through a satellite lens on a clear November day, you might see snow on the tops of our western mountains, in a strip across southern Canada and the Northern Plains, in the lee of the Great Lakes, and along the crests of the northern Appalachian chain. Later, continuous snow cover spreads outward over hills and downward into valleys, onto elevated plateaus, and into lowlands.

Snows and snowpack will reach the shoreline of the Northeast and might visit some northern Pacific beaches in California, Oregon, and Washington once or twice before winter ends. At many locations in Alaska, deep snowpacks occur at tidewater. Coverage always is uneven with regard to latitude, elevation, slope, aspect, and area. There might be snow in southern Kansas and northern Arizona, but none in the chinook zones from south of Denver to as far north as Calgary, Alberta. Often, the West Coast is snowless as far north as southern Canada, while areas east of the Cascades have snow extending south into the Great Basin. On a smaller scale, one commonly finds snow on northerly exposures with none on south-facing slopes. There is less in open wind-swept areas and more in protected locations like lee slopes, gullies, and depressions. Large variation in areal coverage, depth, and snow properties is the norm, but certain wind protected locations, such as Steamboat Springs, Colorado have snowpacks with more uniform characteristics.

In warmer climates south of 40 degrees north latitude, snows usually exist briefly as a single layer from one storm. Such thin deposits repeatedly melt back to the ground throughout winter. Only in colder, snowier regions, does snowpack develop as multi-layered deposits that last for several months. Layers evolve within the snowpack from episodic snowfalls interspersed with other types of precipitation like sleet, rain, or freezing rain, and deposits of surface hoar (frost) and rime. Other layers may be added from the work of wind, sun, and warm temperatures to form crusts that join the mix at random to form a complex, ever-changing structure. In cold conditions, crusts and ice layers within the snowpack disintegrate and become weak, crumbly layers composed of cohesionless grains.

Snowpack may appear as a uniform soft blanket covering the earth, but to a keen observer there is much to see, feel, listen to, and ponder. From the day the snow first covers the ground to the night before it finally melts and disappears, walk or ski through the snow each day and you surely will see what we mean. At times the snowpack is soft. Your steps make little sound and you sink in deeply. At temperatures near zero Fahrenheit, snow crunches and begins to squeak under foot, while at -20°F it almost whistles. Snow also emits booming sounds and wumps and high-pitched swishing noises that indicate catastrophic failure. If you happen to be on a steep slope when this happens, grab a tree or seek more gentle slopes immediately to avoid being caught in an avalanche.

figure 10 A visible view from **NASA LANDSAT** November 24, 1984, showing early winter snowpack in the mountains near Salt Lake City, Utah.

Photo by NOAA/NESDIS/NCDC Asheville, North Carolina

There are times when snow will support most or all your weight. It may crack and crumble some distance away from where you step and each step may leave a slight depression many times larger than your foot. There also will be days when the snow simply crumbles as you walk and falls back into the depression where you stepped. And there are days when you might start an avalanche several yards distant while you still are on nearly flat ground. Beware. How the snow behaves while on the ground is a field of science all its own. We only can touch on it here.

Whole snow crystals that reach the snow surface, and those that arrive splintered and broken, are subject to a process of change called metamorphism. All newly fallen forms undergo a physical transformation from their original delicate structure toward more rounded forms. Points and branchlets progressively disappear with crystal centers gaining mass sublimated from nearby crystal tips, broken arms, and shards. Vapor flows from the smaller particles to the larger ones because the vapor pressure is higher over convexities than it is over flat surfaces and higher over flat surfaces than over concavities. The rate of change is most rapid at warmer snow temperatures – 15 to 32°F – because saturation vapor pressure, with respect to ice, increases with temperature (Figure 1). The process is sometimes called "destructive metamorphism" because the original crystal forms are destroyed. The main result is simplified form and decreased grain size. The smallest recognizable sub-unit in the snowpack is a grain. Grains may be single crystals or might contain several crystals. Deep new snow layers make the transformation from

0 days

3 days

16 days

40 days

figure 11

Metamorphism of a stellar snow crystal over time.

figure 12

Changes of snowpack density over time at selected locations in the United States. Density increases over time within the snowpack as crystal structure decays.

18

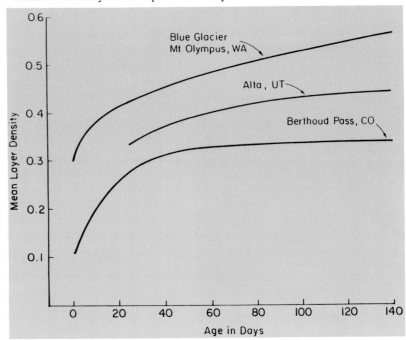

Blue Glacier
Mt. Olympus, WA

Alta, UT

Berthoud Pass, CO

Mean Layer Density

Age in Days

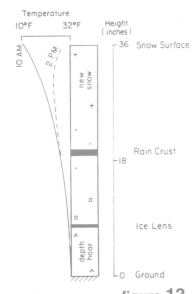

figure 13

Internal snowpack temperature profile at Steamboat Springs, Colorado on February 12, 1971. The total snowpack water content was 7.9 inches. Cohesionless depth hoar (sugar snow) was observed below the first ice lens. Snowpit by A. Judson.

new crystals to older rounded grains more rapidly than do shallow layers because pressure from above accelerates metamorphism. The visible evidence of metamorphism within the snowpack is settling, decreasing snow depth, and increasing density. In the day or two after deep snowfalls, total snow depth decreases by rapid settlement even when new snow is falling, a perplexing phenomenon for ski area marketing staffs.

An important feature of snowpacks is the temperature within the snow. As the snowpack deepens, the insulating properties of snow shield the ground from the severe cold and rapid temperature changes in the air above. This, along with the small heat energy provided by the soil, creates a temperature of about 32°F where the snow and ground meet. Ground surface temperatures less than 32°F occur under shallow snowpacks, but if a deep, fresh snow falls on top, the temperature below will warm again to near freezing. In general, the upper snow layers become much colder than lower ones, especially on clear nights and during cold snaps. Internal snow temperature profiles are monitored by avalanche specialists because temperature influences snow texture and strength.

When the temperature difference or gradient between upper and lower layers rises above about 6°F per foot, vapor from the warm lower layers rises rapidly and is deposited on the colder grains above. This causes another type of metamorphism or change within the snowpack resulting in grains with angular facets. At full development they become striated and appear as hollow cups. These often cohesionless crystals are called depth hoar. Depth hoar can develop overnight in thin snow layers next to crusts, but can take about 12 days or more (under Colorado conditions) to form in snow a foot or more thick. The formation of depth hoar has significant consequences. Snowpacks that were once firm and easy to snowshoe, snowmobile, or ski over now crumble and collapse and can be dangerous.

Snowpack characteristics reflect regional and local climate. In northern Alaska, for example, snowpacks are fairly shallow (about 8-40 inches), wind packed with extreme temperature gradients and full of depth hoar[4]. Astride the Cascades and Sierra lay deep (10-20 foot) snowpacks with minimal temperature gradients. Snowpacks in the Rockies have many distinct layers, while shallower snowpacks in the Northeast often contain mid-winter ice layers and wind pack.

Depth hoar crystals like these form when water vapor from warmer portions of the snowpack rises and is deposited on colder snow grains above.

Photo by Richard L. Armstrong, University of Colorado

Melting Snow

An integral part of the hydrologic cycle is melting snow. It brings sadness to many a snow lover. The inevitable mud season is the good housekeeper's nightmare. Yet, in its own way, the process that takes snow crystals and turns them back into water can be as fascinating as snow itself. All it requires to melt the snow is temperatures above the freezing point and a means of delivering that warmth to the snowpack.

A snowpack melts mainly from the top down. Over the course of a winter, a small amount of heat from the soil below will hold the temperature at the soil-snow interface near 32°F and will melt perhaps a half-inch of snow-water content from beneath the winter snowpack. Much more heat is exchanged at the snow surface. Heat is delivered to the snow surface by solar and terrestrial radiation, rain, snow, condensation of water vapor, and by the direct exchange of heat from the air called "conduction."

There are complex factors that influence the melting of snow.

Photo by Daniel B. Glanz

Many assume that air temperature and sunshine are the biggest factors hastening snowmelt. However, the role of wind should not be underestimated. Turbulent motions caused by strong winds are very effective in delivering heat from the air to the snow surface. Sunshine combined with warm, humid winds work together for maximum melt. Sunshine alone is surprisingly ineffective, unless the snow surface no longer is white. In the sooty, coal-burning industrial cities of the not-too-distant past, snow was observed to melt very quickly.

Rain adds heat to snow and often is credited with causing rapid snowmelt. However, rarely is rain the true culprit. Rain alone cannot melt a deep, cold snowpack. When snow is below freezing, rain freezes on the surface forming crusts and melt is insignificant. Large rainfalls percolate uniformly into snowpacks, raising snow temperatures to the melting point. Typically, much of the water is absorbed in the snow like a sponge and actual melt usually is a small percentage of the rain itself. Only when the temperature of the rain is very warm and is accompanied by wind are melt rates significantly increased. Condensation and sublimation (deposition) of water vapor on snow surfaces give off heat during the phase change. This heat source strongly affects the rate of snow melt in the Northeast, the upper Midwest, and from California north through Washington. In these high-humidity areas, vapor from milder air is deposited on the colder snow surface and significantly hastens snowmelt. This process vividly was demonstrated in the eastern U.S. in January, 1996, when a surge of very warm and moist air with strong winds moved northward over the deep snowpack from the "Blizzard of 1996." With dewpoint temperatures close to 60°F, most of the snow literally disappeared over night. On the other hand, evaporation during snowmelt cools the snow surface and retards snowmelt. This partially explains why snow can melt rapidly at air temperatures just barely above freezing in humid areas, while in the dry areas of the High Plains and Great Basin, snows may linger for days despite daytime air temperatures 10-20°F above freezing.

As melting progresses, much of the water remains in the snowpack. When the snowpack finally contains 3 to 20 percent liquid water by weight, depending on the type and crystal structure of the snow, it reaches its holding capacity and melt water begins to flow. The 3 percent referred to above is typical for old, wet snow and 20 percent liquid water by weight is limited to new, wet snowfall, but only briefly.

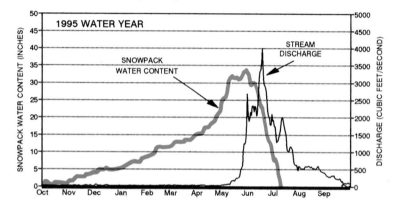

figure 14

Daily snow water content in the upper basin and associated lower-basin stream discharge on the Cache La Poudre River in northern Colorado during 1995. Snowpack water content reached maximum values much later than average due to a very snowy spring resulting in later and greater than normal discharges. Data provided by the USDA Natural Resources Conservation Service.

How fast can snow melt? There is a documented extreme case in Alaska[11] where a chinook wind at about 50°F melted snow containing eight inches of water in two days. Snowmelt models suggest that snowmelt rates of two to three inches of water content per day are possible. Melting rates in the mid-Atlantic states in January 1996 approached two inches of water content per day (12-18 inches of actual snow depth). More typical melt rates during warm, spring weather are around one inch of water per day. For practical applications, snow depths of six inches or less can melt in one day.

Brief Primer on Avalanches

Avalanches are sudden and sometimes frightening demonstrations of the power of moving snow. Most avalanches occur on steep slopes of 25 to 60 degrees although high latitude slush avalanches may occur on slope gradients down to five degrees. The hazard is most pronounced in the western United States, but extends across a tier of northern states into New England and Maine. Avalanches have killed 914 people in the United States in the century ending in 1995. One-hundred years ago, avalanches killed people mainly where they lived and worked because miners built bunkhouses and mine buildings on steep slopes. As the price of silver fell, miners left the mountains and fatalities dwindled. A half-century later, people returned to the snowy slopes in cars and buses, on skis, snowshoes, and snowboards, and once again, became targets for swift, moving snow. Modern day avalanche fatalities slowly are climbing. The steadily rising fatality trend is more uniform, but comes in smaller increments than during the mining days. The increase in fatalities persists despite good avalanche warning programs, avalanche schools, and on the ground training available across the west. Cuts in federal and state avalanche programs are part of the problem as is advertising involving expert skiers and snowboarders "catching air" and landing on super steep slopes. The last full-time professional avalanche research effort in America terminated in 1984.

figure 15

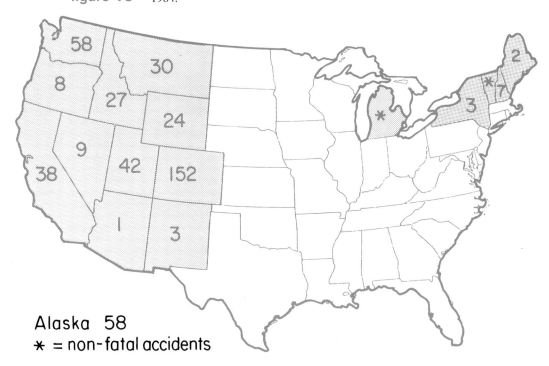

Alaska 58
✱ = non-fatal accidents

Avalanche deaths in the United States by state, for 45 winter seasons, 1950-51 through 1994-95.

Avalanches are classified on the basis of the mode of failure. Those which start at a point and slowly spread in width on descent are called loose-snow avalanches. These avalanches generally involve small amounts of snow with little or no cohesion. Avalanches that start from a fracture line cover a much larger area. These are called slab avalanches and involve cohesive (strongly bonded) snow. The point of departure consists of a smooth and vertical wall of snow of varying width at the upper extension of a sloping-bed surface. The fracture zone of such slides typically is on the order of 10 to 100 feet wide, although such failures may involve snow a mile or more in width. Depth of

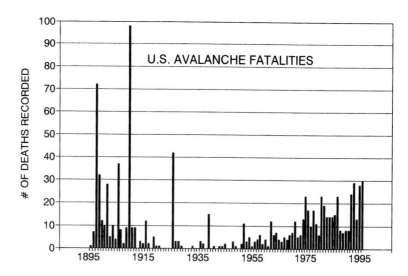

figure 16

U.S. avalanche fatalities, 1896-
1996. Data obtained from the
U.S. Forest Service, the
Colorado Avalanche
Information Center, and other
reliable sources.

Photo by Richard L. Armstrong, University of Colorado

release, usually limited to the newer snow layers, may reach all the way to the ground. The snow in the avalanche area may be wet or dry, hard or soft. It moves downhill at varying speeds – as slow as a brisk walk for wet snow and near 40 to 100 mph or faster for dry snow. Avalanches are generated from weaknesses in the snowpack. Under certain conditions, as little as four to six inches of new snow may initiate dangerous slides.

Avalanche forecasters presently are unable to predict the time of release of individual snowslides, but they, like those who forecast thunderstorms, know the necessary conditions of formation and can warn of expected activity in advance. Despite decades of research in several alpine countries, forecast models are not used routinely in avalanche prediction. On the ground, experienced avalanche personnel reduce the hazard with structures, explosives, and test skiing. To learn more about avalanches, read the avalanche handbooks listed on pages 79 and 80.

A dry snow powder avalanche released by artillery along a highway in southwestern Colorado. Avalanches are a spectacular show of nature and deserve the greatest of respect.

Climatology of Snow in the United States

Photo by Ken Dewey

Photo by Grant Goodge

Snowfall in the United States is an exciting element of our climate and serves both as a valued natural resource and as a natural hazard worthy of great respect. The United States is pounded by storms every year. News headlines show major cities crippled by snow several times each winter. Blizzards sweep the Great Plains from Montana and the Dakotas southward to Texas. Huge dumps of dense, wet snow clobber the Northeastern and mid-Atlantic States almost every year, beautifully decorating the eastern forests, but totally interrupting "life as usual." Meanwhile, out west, hundreds of inches of snow fall over the course of a year in many mountainous areas.

As often as storms seem to hit, statistics on snowfall in the United States show that snow is a brief and often fleeting phenomenon for most densely populated areas. Over much of the southern half of the United States, it may snow only a few hours in an entire year. In the deep South it may snow only a few times in a decade. In Florida, south Texas and southern California, residents may see only a few flakes in a lifetime. Even to the north, where snow falls faithfully every year, a surprisingly small number of storm systems are responsible for most snow accumulation. These few storms, however, can have far-reaching effects.

Storm Tracks

From early autumn into spring, storm systems develop and cross North America heading eastward. Each of these storms is a potential snowmaker if moisture is present and temperatures get cold enough. Areas like eastern Montana and North Dakota are cold enough for snow many months of the year, but they are too far from moisture sources like the Pacific Ocean, Gulf of Mexico, or the Great Lakes to get frequent heavy snows. Significant snowfalls

figure 17

Probability (in percent) of
receiving measurable snowfall
during a winter season. These
figures based on 1930 through
1994 data.

occur in these regions only when organized storm systems help pull moist air
into the area and then lift it higher into the atmosphere causing cooling and
condensation. Seattle and New York City, on the other hand, have plentiful
moisture and frequent organized storm systems, but in these areas the warmer
ocean nearby keeps air temperatures near the ground too warm for snow most
of the time.

During recent years in the United States, there have been an average of 105
storm systems per year that produced snow. The 1973-74 winter season
generated 137 snow-producing storms. Only a handful of years sees less than
85 storm systems cross the country that drop snow somewhere along their
path. Roughly 15 percent of the storms produce heavy snow accumulations of
six to 12 inches or greater.

The snow season for the contiguous 48 United States typically begins in
early October and lasts into late April. At any given location, the date of the
first snow of the season varies a great deal from one year to the next.
Climatologically, latitude and elevation are the primary controls of the dates of
the earliest and latest snows, but individual weather systems and storm patterns
control each year's snow dates. At high elevations and near the Canadian
border, the first snow of the season can come as early as early September and
the last snow in May or June. More than once, snow on the 4th of July has
chilled vacationers in the Rocky Mountains. Occasional August snows send
campers in the northern Rockies scurrying home.

Storms seem to follow similar tracks. While there are no superhighways for
snowstorms that direct their precise paths, there are some preferred
development areas and some common storm tracks. Many low pressure areas

form in the lee of the Rocky Mountains and then progress eastward or northeastward. Some meteorologists refer to storms forming near the panhandles of Texas and Oklahoma as "panhandle lows" that often bring snow to the Midwest. Likewise, many strong storms form along the east coast between the cold continental air masses and the warm maritime air over the Gulf Stream. These storms may be called "Nor'easters" and typically track northeastward along the coast and can affect millions of United States citizens in a matter of just a few hours. There are other storm tracks that have been given special local names. "Alberta Clippers" are fast-moving storms that drop quickly down from Canada and typically bring just a few inches of snow to the Upper Midwest. "Four-corners Lows" are slow-moving storms that develop over the Great Basin and southern Rockies and often bring heavy snow to the High Plains and along the Front Range of the Rockies. The term "Pineapple Express" is used to describe winter storms that sweep into California with copious moisture from the subtropical Pacific.

figure 18

Common tracks and development areas (double circles) for snow-producing storm systems affecting the contiguous 48 United States.

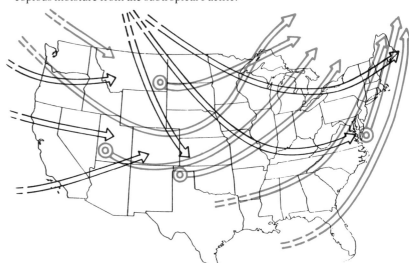

Snow Frequency

Snow falls somewhere in the contiguous 48 States as many as 200 days per year. At individual U.S. weather stations, the number of days with measurable snowfall (0.1 inches or more) ranges from zero to more than 130. The highest frequencies are found in the higher elevations of the Cascade Mountains and the northern and central Rocky Mountains. Frequencies also are high in the immediate lee of the Great Lakes and in northern New England. Snow is most frequent and widespread in January and February.

Climatologists at the National Climatic Data Center, who assisted in the preparation of this book, evaluated snow data for all U.S. weather stations, including Alaska and Hawaii. Over 2,900 weather stations were identified with reasonably complete snowfall data for the period 1961-1990. These stations became the source of data for most analyses in this chapter. Out of these 2,900 stations, 71 percent average measurable snowfall on less than 20 days per year.

figure 19

Percent of U.S. weather stations receiving measurable snowfall (0.1 inches or more) on the number of days indicated (based on 1961-1990 data). Numbers after each bar are the actual number of stations in each category.

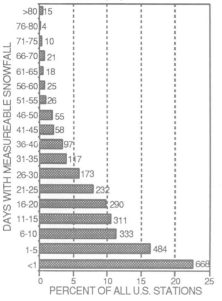

Only 2.3 percent of the stations average more than 60 days with snowfall in the winter season. Most of these stations are located in Alaska, the western mountains, northern New England and the upper Great Lakes. It is helpful to recognize that the density of weather stations in the United States tends to be lower in those areas that receive the most snow.

The Paradise weather station near Mt. Rainier in Washington, elevation 5,427 feet, is the snowiest reporting station in the United States and averages nearly 700 inches of snowfall and 127 days of measurable snowfall annually. The Berthoud Pass weather station in Colorado is located at 11,310 feet and has the same frequency of snowfall, but being further removed from direct maritime moisture sources averages only 390 inches annually. Alta, Utah snowfall observations for the November-April "midwinter" snow season have averaged 481 inches, impressive for a location so far inland.

Figure 20 reveals some interesting aspects of snow climatology in the United States. There is a close association between the number of days with snowfall and total seasonal accumulation. The national average is approximately two inches of snowfall per day on days when snow occurs. Areas along the East and West coasts and across the southern states tend to receive larger snowfall totals on days when snow occurs. Areas in the northern plains and near the Great Lakes, but outside the traditional snowbelt locations, tend to have numerous days with small amounts of snow. Frequent light snows are especially common in northern and central Alaska. Snowfall frequencies in these areas are reasonably high, 60 to 100 days per year, but accumulated snowfall is generally less than 70 inches and results in an average of less than one inch per day of measurable snow. Many of the highest snow averages are found in the mountains of the West and Southwest. Colorado's Wolf Creek Pass averages nearly seven inches of snow per snow day.

figure 20

Scattergraph of average annual snowfall (vertical axis) versus number of days with measurable snowfall (horizontal axis) based on 1961-1990 snow data for approximately 2,900 weather stations in the United States.

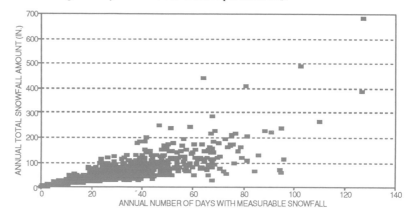

table 1

Snowiest official National Weather Service weather station in each state based on stations with complete or nearly complete snowfall records, 1961-1990. Snowier locations may exist in most states.

Snowiest U.S. Weather Stations

Some of the snowiest places in the world are found in North America. The mountain ranges that guard the west coast of the continent from Alaska southward to northern California include most of the snowiest locations. But every State has its snowiest place. Weather stations may not be maintained in all of the snowiest areas, but perhaps that will change in the years to come.

State	Snowiest Weather Station	Avg Annual Snowfall (in)	Avg # Days with Snowfall > 0.1"
Alabama	Valley Head	6.4	4.1
Alaska	Annex Creek	236.5	51.4
Arizona	Flagstaff WSO AP	109.1	37.8
Arkansas	Harrison FAA AP	14.5	6.8
California	Bowman Dam	248.5	46.8
Colorado	Wolf Creek Pass 1E	441.6	63.9
Connecticut	Norfolk 2SW	98.8	54.0
Delaware	Wilmington WSO AP	21.5	13.5
Florida	Pensacola FAA AP	0.2	0.2
Georgia	Blairsville Exp Station	6.1	3.2
Hawaii	All Stations *	0.0	0.0
Idaho	Island Park	212.9	75.2
Illinois	Antioch 2NW	47.7	26.9
Indiana	South Bend WSO AP	82.8	54.6
Iowa	Dubuque WSO AP	43.3	31.0
Kansas	McDonald	43.0	18.5
Kentucky	Covington WSO AP	24.2	22.7
Louisiana	Plain Dealing	2.2	1.4
Maine	Caribou WSO AP	117.1	65.3
Maryland	Oakland 1SE	83.6	43.9
Massachusetts	Cummington Hill	76.9	26.6
Michigan	Houghton FAA AP	221.4	90.8
Minnesota	Duluth WSO AP	59.9	79.2
Mississippi	Hickory Flat	4.3	2.3
Missouri	Waynesville 2W	27.0	9.8
Montana	Bozeman 12NE	235.8	94.8
Nebraska	Harrison	60.2	23.7
Nevada	Austin	82.1	41.3
New Hampshire	Pinkham Notch	146.0	55.5
New Jersey	Charlotteburg Reservoir	39.5	16.0
New Mexico	Red River	151.2	47.6
New York	Boonville 2SSW	227.2	88.3
North Carolina	Banner Elk	49.0	21.8
North Dakota	Pretty Rock	46.7	29.0
Ohio	Chardon	102.2	45.6
Oklahoma	Boise City 2E	25.4	11.4
Oregon	Crater Lake NPS	488.5	102.2
Pennsylvania	Corry	137.1	59.4
Rhode Island	Providence WSO AP	39.0	20.9
South Carolina	Greenville-Spartburg WSO AP	6.4	3.5
South Dakota	Lead	162.2	66.4
Tennessee	Allardt	20.5	10.6
Texas	Borger	23.8	12.6
Utah	Silver Lake Brighton	407.0	80.5
Vermont	Waterbury 2SSE	121.5	63.0
Virginia	Big Meadows	47.1	17.1
Washington	Rainier Paradise	683.0	127.3
West Virginia	Pickens	160.2	49.8
Wisconsin	Gurney	139.1	59.8
Wyoming	Moose	177.1	70.1

* Snowfalls on higher mountain peaks

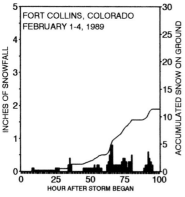

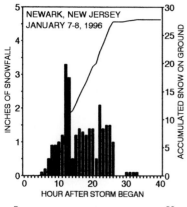

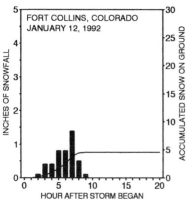

Hourly snowfall totals and accumulations for different types of snow storms: (top) long-duration, frigid-temperature, "upslope" snow storm along the Colorado Front Range; (middle) intense East Coast snow storm; (bottom) typical modest snow storm.

figure 21

Snowfall Duration and Intensity

On average, individual snow events at a particular location last several hours. Prolonged snow events are rare for most parts of the country, except downwind of the Great Lakes and in the mountainous west. Nearly continuous snows in the Olympic Mountains, west of Seattle, are known to fall for periods longer than seven days. Parts of Alaska also are prone to long-duration snows.

Most snow that falls in the United States falls gently and accumulates gradually. Data suggest that more than 90 percent of the time when snow is falling, accumulation rates are less than 0.5 inches per hour. Many of the more intense and disruptive storms, while relatively infrequent, drop snow at much greater rates. Any snow accumulation rate of one inch or more per hour is heavy snow that quickly covers roads and severely limits visibility. Snowfall rates of two inches or more per hour frequently bring travel to a standstill. Such rates are normally short lived and are associated with convective snows or very intense storm systems. These storms may be accompanied by lightning and thunder. Isolated snow accumulations of five to seven inches per hour associated with lake-effect snows in the Great Lakes snow belts have been reported. Extreme storms in the mountainous west and localized convective snow bursts elsewhere in the country have produced similar snowfall rates. Higher accumulation rates up to 18 inches per hour have been reported, but most likely have included drifting snow along with actual snowfall.

Very few weather stations have records of hourly snowfall to document how often snow of various rates actually falls. However, many stations measure hourly precipitation. Hourly precipitation rates during periods of snow at Denver, Colorado's former airport and official weather station, Stapleton Field, show that a trace or more of snow fell in 5,236 hours during a ten-year period (524 hours per year). The water content was either a trace or 0.01 inches in 4,293 of those hours (82 percent). Of the remaining 943 hours, 752 had only 0.02 - 0.05 inches of water equivalent. Hourly water content of falling snow exceeded 0.05 inches in only 191 hours, 3.6 percent of all hours with snow, and equalled or exceeded 0.10 inches in just 70 hours. Many of the worst problems associated with transportation and snow removal result from the relatively few hours of high intensity, high water content snowfall.

figure 22

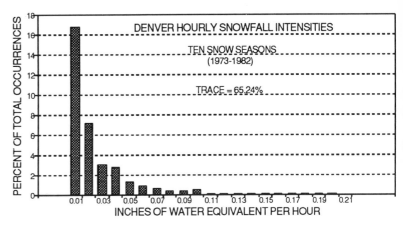

DENVER HOURLY SNOWFALL INTENSITIES

TEN SNOW SEASONS
(1973-1982)

TRACE = 65.24%

PERCENT OF TOTAL OCCURRENCES (y-axis: 0 to 18)

INCHES OF WATER EQUIVALENT PER HOUR (x-axis: 0.01 to 0.21)

figure 23

Snowfall patterns from selected storms: a) March 24, 1983, southeastern U.S. storm, b) Nov. 26-28, 1983, over the Northern Plains, and c) Feb. 10-12, 1983, East Coast snowstorm.

Hourly precipitation amounts for all snow-only events for Denver, Colorado, 1973-1982. This distribution may vary from region to region.

At a point, storms may last only a few hours. But the storm systems themselves often take several days to cross the country and may drop snow over huge areas. Fall and spring storms tend to move more slowly than mid-winter storms. However, due to warmer temperatures these storms also are more likely to produce primarily rain and only drop snow over small areas or at high elevations. Mid-winter storms tend to progress swiftly across the country from west to east, but may drop snow along their entire track.

East of the Rocky Mountains, storms tend to deposit snow in irregular or very smooth bands. In mountainous areas, the topography often is the dominant factor and contributes to much more irregular snowfall distributions.

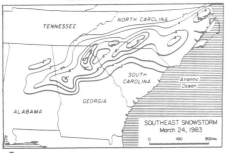

A

SOUTHEAST SNOWSTORM
March 24, 1983

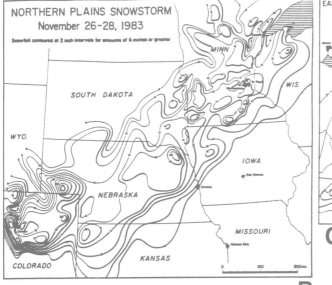

NORTHERN PLAINS SNOWSTORM
November 26-28, 1983

Snowfall contoured at 2 inch intervals for amounts of 6 inches or greater

B

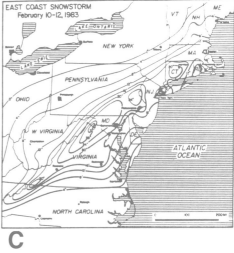

EAST COAST SNOWSTORM
February 10-12, 1983

C

Snowfall Seasonality

Temperature is the primary control of the seasonality of snowfall. Nationwide, about 70 percent of annual snowfall at designated weather stations falls from December through February. January is the snowiest month of the year for about half of the U.S. But other factors also affect snowfall. In the northern Great Lakes region, warm lake water early in the winter season enhances early snowfall by adding extra moisture and convective instability to cold air masses. In the High Plains and along the eastern foothills of the Rockies the air is too dry in mid-winter to allow much snowfall. In those areas, more snow falls in spring and autumn than in mid-winter due to a combination of greater available moisture and more storms that produce easterly "upslope" winds. Stations in northern Alaska are, perhaps, most unique. There, a large percentage of the season's snow falls before the end of November. During the intense cold of mid-winter, very light snows occur frequently, but with minimal accumulation.

figure 24

Average monthly snowfall totals (inches) for selected locations in the United States based on 1961-1990 data showing differences in seasonal snowfall patterns across the country.

For many applications, the frequency and probability of snowfall is more valuable information than total accumulated snowfall. Knowledge about how many days to expect with certain amounts of snowfall can be very helpful for planning ahead, if your business is affected by snow. Graphs similar to Figure 25, which show how many snows equal to or greater than specified amounts to expect over an entire winter, can be produced for any part of the country where many years of accurate daily snowfall data are available.

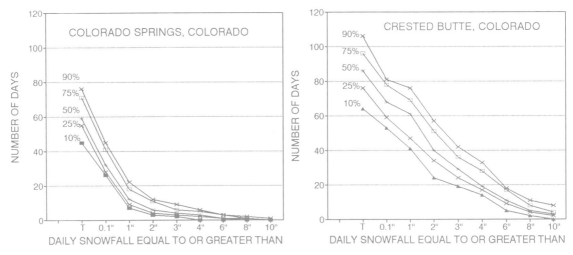

figure 25

Snow planning graphs. These figures show the specified number of days receiving equal to or greater than the specified amount of daily snowfall during an entire winter season for selected probabilities of occurrence for Colorado Springs and Crested Butte, Colorado.

National Snowfall Patterns

The information on the following pages provide a descriptive comparison of snowfall totals, snowfall frequencies, and snow cover duration for the United States. These summaries are based on snowfall data for the period 1961-1990 for more than 2,900 reporting stations with complete or nearly complete data.

There are two basic snowfall regimes over the United States. 1) East of the Rocky Mountains, snowfall increases systematically from south to north. The Appalachian Mountains, the Great Lakes and the Atlantic Ocean all influence this pattern, but temperature and latitude are the primary controls. 2) From the Rocky Mountains to the Pacific Ocean, the primary controls for snowfall are elevation and orography. Snowfall increases with elevation. However, for a given elevation, more snow usually falls on the upwind side (west-facing) of mountain ranges than the downwind (east-facing) side. Average annual snowfall, snowfall frequencies and snow-cover duration all vary dramatically over very short distances in the West, producing very complicated patterns on these maps. Many of the largest cities in the West are located in relatively snowfree locations, but a trip of 50 miles or less may take you to areas receiving several-hundred inches of snow.

table 2

table 2 **Comparative snowfall data for selected U.S. locations. Averages are based on 1961-1990 data. Extremes are derived from the entire period of observation.**

	Mean Annual Snowfall (inches)	% Annual Snowfall Falling in Dec.-Feb.	Snowiest Month	Mean Annual # of Days			Greatest Daily Snowfall (inches)		Greatest Daily Snowdepth (inches)		Snowiest Season***	
				Snowfall >1.0"	Snowfall >5.0"	Depth >1"	Amount	Date	Amount	Date	Amount	Season
Albuquerque, NM	11.4	68	JAN	4.1	0.1	6.8	14.2	12/28-29/58	14	12/29/58	37.4	1972-73
Anchorage, AK	68.5	50	DEC	20.9	2.6	150.7	17.7	12/28-29/55	47	12/30/55	132.6	1954-55
Asheville, NC	16.0	75	JAN	4.6	0.5	10.7	18.5	3/12-13/93	20	3/14/93	48.2	1968-69
Atlanta, GA	2.3	78	JAN	0.7	*	1.1	8.3	1/23/40	8	1/23/40	13.2	1894-95
Baltimore, MD	22.3	81	FEB	5.7	1.1	15.5	24.5	1/28/22	31	2/13/1899	62.5	1995-96
Birmingham, AL	1.5	80	JAN	0.4	*	0.7	13.0	3/12-13/93	13	3/14/93	13.0	1992-93
Bismarck, ND	43.2	52	MAR	13.4	1.1	90.6	15.5	3/3/66	28	2/23/79	91.8	1993-94
Boise, ID	21.6	75	DEC	7.6	0.2	28.4	17.0	12/16-17/1884	22	12/17/1884	50.0	1916-17
Boston, MA	42.7	77	FEB	11.1	2.3	38.4	23.6	2/6-7/78	32	1/10/96	107.6	1995-96
Burlington, VT	82.9	68	DEC	23.5	3.1	94.9	24.2	1/13-14/34	33	12/28/69+	145.4	1970-71
Casper, WY	82.4	40	MAR	25.4	3.0	68.8	31.1	12/23-24/82	21	12/24/82	151.6	1982-83
Chicago, IL	38.8	72	JAN	11.9	1.1	46.6	18.1	1/26-27/67	29	1/14/79	89.7	1978-79
Columbus, OH	29.8	75	JAN	9.7	0.7	29.9	12.3	4/4/87	17	1/23/78	67.8	1909-10
Crater Lake, OR	495.0	49	MAR	93.0	39.6	240.0	37.0	2/28/71	252	4/3/83	879.0	1932-33
Denver, CO	60.7	38	MAR	17.7	2.7	49.3	23.6	12/24/82	33	12/6/13	118.7	1908-09
Detroit, MI	42.6	73	DEC	13.7	1.3	52.0	24.5	4/6/1886	26	3/5/00	78.0	1925-26
Dubuque, IA	43.3	65	DEC	13.6	1.4	70.8	15.5	3/4-5/59	27	2/9/1893	75.7	1961-62
Elkins, WV	75.5	71	JAN	24.3	2.6	50.3	18.8	1/7-8/96	20	1/26/77	136.6	1995-96
Flagstaff, AZ	109.1	53	MAR	23.7	7.3	71.1	27.3	12/13/67	68	1/25/49	210.0	1972-73
Ft. Worth, TX	3.0	87	JAN	1.2	*	2.2	12.1	1/15-16/64	8	1/16/64+	17.6	1977-78
Great Falls, MT	63.3	44	MAR	20.6	1.8	65.2	16.8	4/20/73	24	4/9/75	117.5	1988-89
Green Bay, WI	49.1	67	DEC	15.3	1.6	90.4	15.3	1/26-27/96	25	1/25/79	79.6	1922-23
Hartford, CT	49.1	77	JAN	12.3	2.7	56.3	21.0	2/11-12/83	38	1/13/96	114.6**	1995-96
Houghton, MI	221.4	72	JAN	59.7	12.6	151.0	24.9	3/4/85	72	3/1/37	376.1	1978-79
Little Rock, AR	5.8	83	JAN	2.1	0.2	9.1	13.0	1/2/1893	13	1/2/1893	26.6	1959-60
Louisville, KY	17.5	75	JAN	4.9	0.6	14.9	15.9	1/17/94	19	1/20/78	50.2	1917-18
Lynchburg, VA	21.8	79	FEB	5.8	1.2	15.6	19.0	1/6-7/96	25	1/30/66	56.8	1995-96
Minneapolis, MN	57.4	59	JAN	17.5	2.1	96.7	21.0	10/31-11/1/91	38	1/23/82	98.4	1983-84
Nashville, TN	11.0	81	JAN	3.6	0.4	7.7	17.0	3/17/1892	17	3/17/1892	38.5	1959-60
New York, NY	24.3	83	FEB	7.0	1.1	22.7	26.4	12/26-27/47	26	12/27/47	75.6	1995-96
Omaha, NE	28.7	67	JAN	9.5	0.9	46.8	18.3	2/11/65	27	3/16/60	67.5	1911-12
Philadelphia, PA	23.7	81	JAN	6.7	0.9	17.4	27.6**	1/6-7/96	28	1/7/96	65.5**	1995-96
Pittsburgh, PA	44.8	71	JAN	13.8	1.1	43.0	23.8	3/13/93	26	1/22/78	82.0	1950-51
Portland, ME	72.5	73	JAN	17.5	4.0	85.3	27.1	1/17-18/79	55	1/17/23	125.5	1922-23
Portland, OR	6.6	91	JAN	2.0	0.3	2.7	16.0	1/31-2/1/37	19	2/6/1893	50.0	1871-72
Providence, RI	39.0	78	JAN	10.2	2.1	35.9	27.6	2/6-7/78	30	2/5/61	106.1	1995-96
Rapid City, SD	40.4	42	MAR	12.7	0.8	58.9	18.3	3/31-4/1/27	17	4/19/70+	80.9	1985-86
Reno, NV	25.5	60	JAN	7.9	0.9	16.5	22.5	1/17/16	30	1/18/16	72.3	1915-16
Salt Lake City, UT	64.6	56	DEC	19.3	2.6	52.1	18.4	10/17-18/84	23	1/31/42	117.3	1951-52
Seattle-Tacoma, WA	10.9	82	JAN	3.4	0.5	5.3	21.5	2/2/16	29	2/2/16	67.5	1968-69
South Bend, IN	82.8	73	JAN	25.6	3.0	71.6	17.5	11/25-26/77	41	1/30/78	172.0	1977-78
Spokane, WA	48.0	76	DEC	16.4	1.4	51.6	13.0	1/6-7/50	42	2/1/69	93.5	1949-50
St. Louis, MO	22.5	74	JAN	6.2	1.0	23.3	20.4	3/30-31/1890	20	2/11/82+	67.6	1911-12
Syracuse, NY	110.8	74	JAN	33.1	4.1	85.6	35.6	3/13-14/93	48	2/1/66+	192.1	1992-93
Topeka, KS	20.8	78	JAN	6.8	0.7	27.0	18.7	2/27-28/00	18	3/15/60	47.9	1911-12
Tulsa, OK	8.9	79	JAN	3.2	0.2	8.5	12.9	3/8/94	11	1/8/88	28.7	1923-24
Washington, D.C.	23.6	81	JAN	6.0	1.3	20.5	25.0	1/27-28/22	25	1/8/96	54.4	1898-99
Wolf Creek Pass, CO	441.6	53	DEC	62.5	32.2	197.0	42.0	1/5/73	251	3/31/79	837.5	1978-79

+ Last of more than one occurrence
* Less than 0.1 days
** Measurement may not be official

*** Changes in snow measurement policies in 1995-96 have made comparisons difficult.

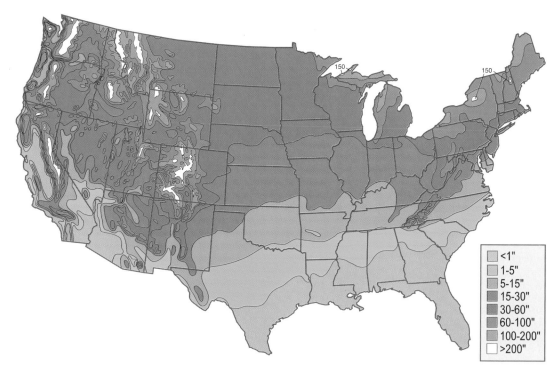

figure 26   Mean annual snowfall for the United States for the period 1961-1990.

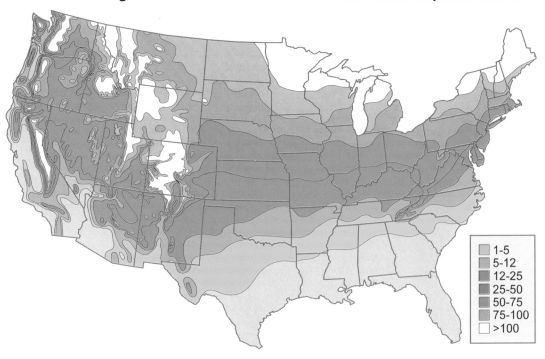

   figure 27   Mean number of days per year with one inch or more of snow on the ground for the 1961-1990 period.

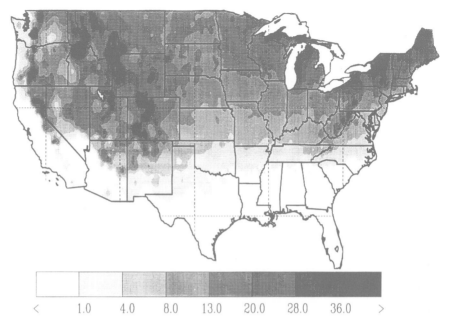

< 1.0 4.0 8.0 13.0 20.0 28.0 36.0 >

figure 28 Mean number of days per year with snowfall of one inch or greater based on data for the 1961-1990 period.

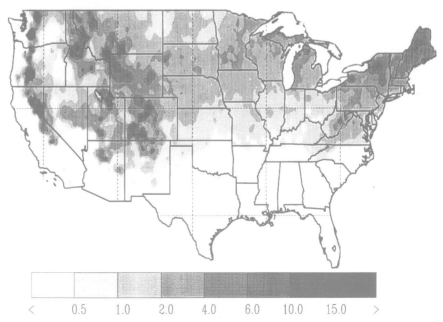

< 0.5 1.0 2.0 4.0 6.0 10.0 15.0 >

figure 29 Mean number of days per year with snowfall of five inches or greater based on data for the 1961-1990 period.

figure 30

Mean annual snowfall for Alaska for the period 1961-1990. Local patterns were estimated using the National Weather Service Alaska snowfall map based on all available data through 1972.

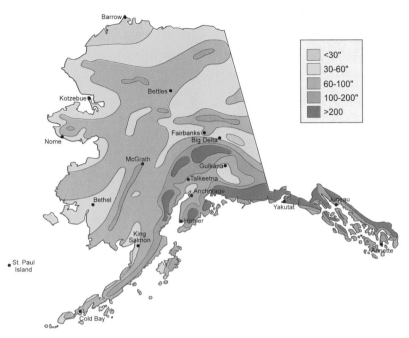

	<30"
	30-60"
	60-100"
	100-200"
	>200

table 3

Station	Mean Annual Snowfall (inches)	Mean Annual Number of Days With:		
		Snowfall > 1.0"	Snowfall > 5.0"	Snowdepth > 1"
Anchorage	68.5	20.9	2.6	151
Annette	51.4	15.3	2.3	26
Barrow	29.1	7.6	*	252
Bethel	46.7	16.1	0.4	161
Bettles	84.3	27.3	2.7	216
Big Delta	46.1	18.2	0.7	188
Cold Bay	67.2	23.0	0.7	71
Fairbanks	71.4	23.3	1.6	191
Gulkana	47.9	17.0	1.3	180
Homer	60.0	19.5	2.3	102
Juneau	105.4	28.5	5.5	78
King Salmon	47.7	16.8	0.8	93
Kotzebue	50.5	18.1	0.1	220
McGrath	98.2	31.0	3.1	201
Nome	55.8	18.4	0.4	188
St Paul Island	61.9	18.9	0.7	120
Talkeetna	124.0	33.4	6.4	193
Yakutat	205.4	48.1	13.5	130

* Less than 0.1 days

Comparative snowfall data for selected Alaskan locations. Averages are based on 1961-1990 data. The locations of these weather stations are shown in Figure 30.

Alaska snow deserves its own book. Outside the mountainous regions of central and southern Alaska, areas of heavy snowfall are more limited than its reputation might suggest. Most coastal locations and interior regions of the state average between 30 and 80 inches of snowfall per year – not unlike the Upper Midwest or southern New England. What is different from the "lower 48" is the higher frequency of light snow and much longer snow cover durations (Table 3).

Snowfall Variability

The extraordinary variability in snowfall from place to place and from year to year is one of the great challenges in planning for snow and adapting to it. The number of days with snow at any location may vary by more than 300 percent from one year to the next. Snowfall amounts may vary even more. It is common to have a year with record snowfall followed by a year with little snow. On occasion, several winters in a row will be severe in the same part of the country like the late 1970s in the Midwest and Northeast.

The seasonality of snowfall also changes greatly from year to year. In any given year, there often are winter months with little or no snowfall, even at relatively snowy places. These may be

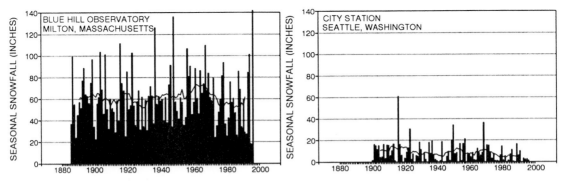

figure 31

Seasonal snowfall totals
through history at Blue Hill
Observatory, Milton,
Massachusetts, and the Seattle,
Washington city weather
station.

followed by stormy periods that produce several times more snowfall than
average. A single snowstorm may contribute the majority of the season total in
less snowy portions of the country.

Spatial patterns of snowfall vary dramatically. Averaged over time,
snowfall patterns become increasingly similar to what is shown in Figure 26,
but in any given year, the patterns can look much different.

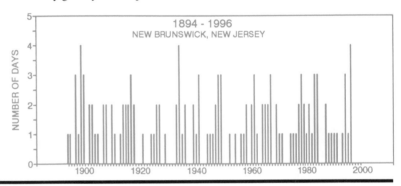

figure 32

The number of days each
winter season, 1894 - 1996 on
which at least five inches of new
snow fell at New Brunswick,
New Jersey. The number of
days with heavy snowfall varies
from year to year with no
predictable pattern.

figure 33

The range of monthly snowfall
totals for the 1961-1990 period
for selected U.S. locations. The
maximum observed monthly
totals are shown along with
totals that have been exceeded
in 25 percent, 50 percent and 75
percent of the years,
respectively.

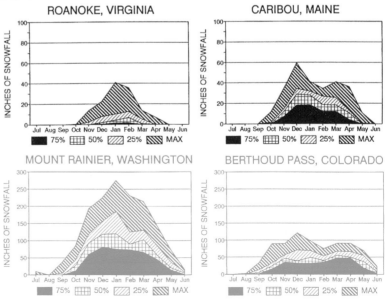

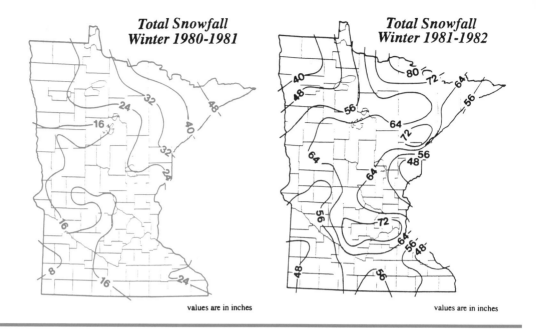

Total Snowfall
Winter 1980-1981

values are in inches   values are in inches

figure 34

Snow Hydrology and Snowloads

Total seasonal snowfall patterns for two successive winters, 1980-81 and 1981-82, for Minnesota. These figures compliments of the Office of the Minnesota State Climatologist.

Snow hydrology, the study of the flow of water from melting snow, is a fascinating topic. Many professionals dedicate their entire careers to studying and predicting how and when rivers and streams will rise and flood with the water from melting snow. The areal extent of snow cover, its water content and melt rates are obviously critical starting points. But other factors like future weather conditions, soil moisture, frozen soils, existing stream flow, the shape, slope and orientation of watersheds, basin geology and land use must be considered when relating snow conditions to potential runoff and flooding.

Over much of the United States, snow falls and melts again within just a few days. But in Alaska, New England, the upper Great Lakes, Northern Plains, Rocky Mountains, Sierra Nevada, and Cascades, snow accumulation and subsequent melt play a dominant role in the hydrologic cycle. Runoff from melting mountain snow contributes 60 to 75 percent to the annual streamflow of many rivers in the Rocky Mountain region. To understand these processes and anticipate water supplies and potential flooding, the measurement of total water content of the snow lying on the ground is extremely important. Figure 35 shows characteristics of the accumulation and melt of snow in different parts of the country.

In Alaska, except in milder coastal areas, in the high mountains of the western United States, in northern New England, and in the far northern portions of the upper Midwest, snow normally accumulates steadily throughout the winter. The water content of snow on the ground may reach a maximum in March or as late as May in the highest mountain ranges of the West. Snowmelt then can occur quickly and result in rapid runoff and high flows on rivers and streams (see Figure 14).

Since the 1930s, the United States Department of Agriculture Natural Resources Conservation Service, formerly the Soil Conservation Service, has routinely measured the water content of snow on the ground in the mountainous West. By knowing the water content of snow on the ground in March, April, and May and soil moisture from the previous autumn, it is possible to predict with reasonable confidence the amount of water in the rivers and reservoirs during the summer season. In the mountains, daytime temperatures decrease with elevation. This results in maximum melt rates that only occur in certain elevation bands on any given day and gradually work their way up the mountains. This tends to limit the potential for rapid and severe flooding. The situation is more challenging in the Upper Midwest and East. Even though these areas get much less total accumulation of snow than high mountain areas, the horizontal extent of snow cover is greater, and wide areas tend to melt at the same time. The January, 1996, snowmelt in the mid-Atlantic states was a dramatic example.

For the majority of the country, where snow comes and goes during the winter, rain is a much bigger threat than melting snow for flooding.

The term "snow load" refers to the weight of snow accumulated on natural or man-made structures. Tragedy can occur if the load, often calculated in pounds per square foot, exceeds the capacity of the structure. Many of us have seen broken tree branches following a heavy, wet snow. However, it often takes a disaster like the 1922 collapse of the Knickerbocker Theater in Washington D.C. to draw attention to the importance of knowing the weight of snow.

For snow loads, just as in hydrologic applications, the depth of snow is not what matters most. It is the total water content of the accumulated snow that produces the weight or load. For example, a fall of 12 inches of snow containing one inch of water weighs 5.2 pounds per square foot. If no melting occurs and the next storm brings an additional 25 inches of snow containing two inches of water (10.4 pounds per square foot), the total snow load would become 15.6 pounds per square foot. If the area of the roof is 1,800 square feet, typical of many homes in the U.S., there would be 28,080 pounds of snow resting on the roof – the weight of three mature elephants. No wonder you sometimes see crews of workers shovelling roofs after periods of heavy snow.

It is tempting to quickly judge the weight of snow based on total snow depth. But always remember that the density of snow can vary greatly. Three feet of fresh snow may contain less than two inches of water. But three feet of aged, compacted snow, or melting wet snow with rain falling onto it could weigh many times more.

figure 35

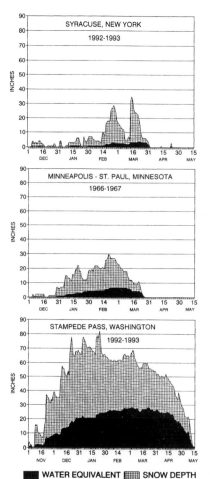

WATER EQUIVALENT ▦ SNOW DEPTH

Daily snow depth and water equivalent of snow on the ground for selected winter seasons at Syracuse, New York, Minneapolis-St. Paul, Minnesota, and Stampede Pass, Washington.

Record Snowstorms

Almost every year in the United States some locations experience a snowstorm that exceeds any previous storm at that location. Each winter there may be five to 20 storms that are remarkable in some way, and that doesn't include Alaska. It is difficult to select the worst storms of record for the United States. There are hundreds to choose from, all of them legitimate "record storms."

This drug store in Weaverville, North Carolina, collapsed under the weight of snow from the March 13, 1993, record snowstorm.

Photo by Grant Goodge.

DENNIS THE MENACE

"OF COURSE, IT'S **NOTHING** LIKE THE SNOWSTORMS WE HAD WHEN I WAS *YOUR* AGE."

What constitutes a record storm is distinctly a function of geographic location. Four inches of snow in the deep South could represent a locally "mammoth" storm, while in Milwaukee, Wisconsin snowfall must exceed 12 inches to rate as extraordinary. In New England and parts of the Appalachian Mountains at least two feet in 24 hours must fall before a storm comes close to a record. For the western mountains and some of the Great Lakes snowbelts, two feet of snow are fairly common. In those areas it is the storm episodes that dump heavy snow for the most number of days or with the coldest temperatures that may be the most memorable.

In the blizzard country of the Great Plains, it is the accompanying wind, temperature, and the size of the drifts that often are compared. Fifteen-inches of level snow may be a minor inconvenience compared to the impacts from six inches pushed by sixty mile-per-hour winds.

75.8"

The maximum 24-hour snowfall in the United States is an astonishing 75.8 inches at Silver Lake (elevation of 10,360 feet in the mountains west of Boulder, Colorado). It occurred April 14-15, 1921. The most snowfall in a 12-month period at an official weather station was 1,122 inches during the 1971-72 winter at the Mount Rainier Paradise Ranger Station. Proper and accurate measurement of these extreme events is one of the great challenges facing weather observers.

In some ways, it may be more useful to count the number of roads, airports and businesses closed. This may be a better indicator of storm severity than snowfall itself. Recovery time also may be a useful concept. Southern cities may take two days to recover from an inch of snow while snowbelt cities and ski resorts may be back to normal in 24 hours after a "two-footer."

If you enjoy reading accounts of historical major United States snowstorms, several books have been written by noted meteorologist and author David Ludlum. There also are many exciting storm accounts in the magazine Ludlum edited for many years, *Weatherwise*.

Selected Large-Area U.S. Storms

Historically there have been relatively few gigantic storms that have brought deep snow and hardship to broad areas of the country at the same time. Here are a few noteworthy examples.

• **January 11-13, 1888.** A large blizzard brought misery and death to expansive areas of the central United States from Texas to Canada.

• **March 2-5, 1960.** A massive storm system dropped heavy snow from the western Midwest to New England and Canada.

• **March 13-14, 1993.** Extremely heavy snows fell from Mississippi to Maine. Howling winds and record cold temperatures brought much of the eastern United States to a standstill.

Snow removal the old-fashioned way following nearly four feet of snow in Denver, Colorado, in December 1913.

Photo from the Colorado Historical Society

This photo of the Kiel family farm in Garden County, Nebraska, shows how the "Blizzard of 1949" literally buried parts of the western Great Plains beneath house-high drifts.

Photo from the Nebraska State Historical Society

Memorable
Regional Storms

There is nothing more stressful than a long drive through heavy snow at night. Sometimes just the faint view of tail lights in front of you is enough to overcome the hypnotic effect of thousands of flakes converging toward your headlights.

Photo by Ken Dewey

Eastern United States

March 11-14, 1888. One of the worst and most popularized historical snowstorms dropped several feet of wind-blown snow across the eastern seaboard from Chesapeake Bay to Maine and paralyzed most major cities. Middletown, Connecticut received 50 inches.

February 11-14, 1899. Two to three feet of snow fell from Washington D.C. to New York City with 44 inches reported in New Jersey. Snow also fell over Florida. Subzero temperatures followed. Even Tallahassee reported -2°F, the coldest temperature on record for Florida.

January 27-29, 1922. An intense storm struck from South Carolina to Massachusetts with a fury of heavy, wet snow. Twenty- eight inches fell in Washington D.C. Numerous buildings collapsed.

April 2-5, 1987. An extremely large snow for so late in the season dropped snow from Alabama and Georgia northward to New York. More than a foot fell throughout the Appalachian Region with 40-60 inches in the Great Smoky Mountains. Rapid snowmelt after the storm produced flooding.

January 6-8, 1996. One to three feet of wind-driven snow buried most of the Atlantic Coast from Virginia to Boston. Record snowfall of 30 inches was reported at Philadelphia.

Deep South

February 14-15, 1895. Snow fell all the way to the Gulf Coast. Galveston, Texas observed 15.4 inches and New Orleans got 8.2 inches. Amazingly, 24 inches in 24 hours accumulated at Rayne, Louisiana.

November 16, 1958. Arizona experienced an early-season storm that dropped a record 6.4 inches at Tucson.

February 8-10, 1973. More than a foot of snow fell from southern Alabama to the beaches of southeastern North Carolina.

Midwest/Great Lakes

November 11-12, 1940. An intense autumn storm clobbered Iowa, Minnesota and the Northern Great Lakes. Over 100 fatalities were noted, many as a result of cold and exposure as hunting season had just begun.

January 28-31, 1977. A true blizzard produced several days of "white-out" conditions over western New York resulting in a federal disaster declaration.

January 26, 1978. A North Dakota-like blizzard with hurricane-force winds, temperatures near zero, and very intense snow paralyzed Ohio, Indiana, southern Michigan and eastern Illinois. Drifts totally covered cars and trucks on major highways. Many lives were lost, some by suffocation and freezing in snowbound vehicles.

December 7-11, 1995. A monstrous lake-effect snowstorm dumped 37.9 inches at Buffalo, New York in 24 hours. Sault Ste. Marie, Michigan, not known for extreme lake-effect snowfall, received 56.5 inches in 72 hours during this storm episode enroute to their snowiest winter on record.

Plains

January 6-13, 1886. A terrible blizzard from Texas to Iowa was especially severe in Kansas where more than 50 people died and an estimated 80 percent of the cattle in the state were killed.

January 1-6, 1949. An extreme blizzard struck the northern plains and northern Rockies. Thirty-nine deaths were reported in Wyoming, Nebraska, and Colorado with snowdrifts totally covering houses and barns.

March 22-25, 1957. The southern plains from New Mexico to Kansas were blasted with extreme winds and snow. Drifts as deep as 30 feet were found in parts of Texas.

March 2-5, 1966. A four-day Dakotas blizzard produced 30-foot drifts and wind gusts that reached 100 mph.

The West

December 1-5, 1913. A huge dump of wet snow fell from New Mexico northward to Wyoming. Many areas got close to four feet on the level with drifts reported to more than 25 feet in open areas. Denver totalled 46 inches for the storm while Georgetown, in the mountains west of Denver, got 86 inches.

January 31 - February 2, 1916. The greatest Seattle snowstorm of all time dropped 32.5 inches on the city. Many buildings collapsed and all transportation was halted.

December 29-31, 1951. An onslaught of Pacific moisture flushed across California and pushed into the central Rockies dropping up to seven feet of snow. Several fatalities resulted from avalanches and exposure, most in Colorado. Transportation was halted for nearly a week.

January 11-16, 1952. Record snows were reported in several locations in California, Nevada, and Idaho. Tahoe, California totalled 149 inches for the storm.

May 28-30, 1982. A late spring storm buried parts of central Montana beneath 60 inches of wet snow with 15-foot drifts. Hundreds of miles of powerlines came down and some areas were without electricity for more than four weeks.

January 1995. A series of storms crashed into California and deposited tremendous amounts of snow in the Sierra Nevada. Snow piled up to depths of 25 feet on the level in portions of the mountains. The weight of the snow damaged many roofs, trees, and other structures.

Incredible Local Storms

April 14-15, 1921. An intense spring storm with copious moisture buried portions of the Colorado Front Range with one to four feet of wet snow, while nearby valleys and plains received mostly rain. A total of 87 inches fell at Silver Lake including a North American record of 75.8 inches in 24 hours.

October 17-18, 1984. A sudden autumn snowstorm along Utah's Wasatch Range locally was enhanced by the Great Salt Lake. Parts of the Salt Lake City area received one to two feet of snow while no snow was reported just 10 miles west of the city.

January 4-5, 1988. A localized lake-effect snowstorm dropped 69.5 inches of snow on the Tug Hill area of New York while areas only a few miles to the north and south got less than six inches.

Huge storms may require massive snow removal efforts. These trucks teamed up to clear the streets in Loveland, Colorado, after 15 inches of snow fell March 6-7, 1990, with a remarkable water content of more than three inches.

Photo by Joel Radtke, *Loveland Reporter Herald*

Measuring Snow

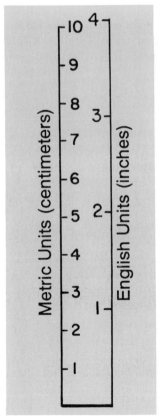

**1 inch = 2.54 centimeters
25.4 millimeters**

We measure snow to help us describe, compare, study and prepare for it. For more than 100 years there has been a nationwide effort to systematically measure and record snow in the United States. While we hope that the snow data in the record books are accurate and consistent, we are never sure. Measuring snow is not easy. It never has been and never will be. Snow is so elusive and dynamic that snow measurements often are uncertain. Under some circumstances, it is possible to send out eight independent weather observers and end up with eight different snowfall reports. Although attempts have been made to standardize observational procedures, not all weather observers or organizations use the same techniques. Some of the maps and graphs in the previous chapter are not as accurate as they could be because of measurement inconsistencies. For many of the nation's worst snowstorms, there is a high degree of uncertainty regarding reported snowfall amounts.

In this day and age, we often turn to technology to help us solve our problems. Technology provides many electronic methods for measuring a variety of weather variables. Some of these methods have greatly improved our ability to observe and understand our climate. For example, data collected from a variety of satellite observing systems provide timely information about cloud patterns, temperatures, winds, water vapor, snow cover, and snow water content across the entire earth. These sources of global data are of great value to both climatologists and weather forecasters, but they also are imperfect and do not replace the need for accurate local surface weather observations.

Automation of surface weather observations began in the 1970s and continues today. Replacement of human weather observers and mechanical weather instruments by electronic monitoring devices linked to computers has taken place rapidly during the 1990s. This change allows more frequent and more remote weather observations and has proven to be quite satisfactory for weather elements like temperature, pressure, and wind. But the accurate electronic measurement of snow continues to elude even the most brilliant engineer. The total accumulated water content of snow on the ground has been measured in the mountainous West since the late 1970s with good success[12]. But cost-effective-automated measurements of snowfall rates, daily snowfall totals, daily (or more frequent) water content, and depth of snow on the ground have been problematic at best. For now, the best source of information continues to be snowfall data carefully gathered by trained and skilled human observers who understand and enjoy the whimsical nature of snow.

Unfortunately, there aren't enough skilled observers who are wise to the tricks of snow. Many studies by climatologists, engineers, and scientists concerned with the properties and impacts of snow are hampered by a lack of accurate and reliable snow data. Robinson (1989)[9], while studying historic snowfall data for the United States, was disappointed to find very few locations across the country with serially complete and accurate measurements.

There are several different aspects to measuring snow. For the purposes of military and civilian aviation, snow measurements may require detailed information gathered hourly or more frequently that include observations of the height and type of clouds, visibility (how far you can see), and changes in the type and intensity of snowfall. For weather forecasters and ground transportation specialists, accumulation rates and times that snow begins, ends, or changes form may be especially important. For hydrologists, the main concerns are precipitation rates, melting rates, and the total water content of all snow on the ground. For climatological purposes, it is important that depth of new and old snow and the water content be reported daily. There are countless climatological uses and applications of snow data for which observations must be consistent over time and also from place to place.

The paragraphs that follow are intended to offer training and assistance to those who take daily climatological observations of snow. Also, for those who must utilize snow data, this should offer insight into how the data may have been collected and what problems may exist.

Historical Perspective

As it is so often and appropriately said, "There is nothing new under the sun." Of course, we think ourselves wiser and more experienced in the difficulties of measuring snow than anyone before us. Most likely, we are wrong. Of the various challenges standing in the way of accurate and consistent observations of snow that we will describe later in this chapter, many were addressed quite thoroughly long ago.

Allow us to use the instructions prepared in the 19th Century by Cleveland Abbe, renowned scientist and meteorological observer, as an introduction to this section on measuring snow.

"For collecting snow a simple cylinder two feet deep is used . . . In case of light dry snow-flakes and high wind, however, all gauges frequently fail to collect or to retain the proper amount of snow, and the catch should, when possible, be checked, by measurement of the snow on an undrifted uniform flat surface. This is best done by plunging the cylinder vertically into the level snow until its lower edge reaches either the ground or the upper surface of the snow that fell since the last measurement. A sheet of tin is then slipped under the gauge; the snow thus collected is melted, and its amount represents the height of the snowfall in equivalent inches of water. . . . The measurement of

A winter view of the official cooperative weather station in Gunnison, Colorado. This station is one of a few hundred in the country with 100 years or more of snow observations.

Photo by Nolan Doesken

the height of water by assuming ten inches of snow to one of water, or some other factor, is subject to a large range of error because of the wide variability of this ratio for different kinds of snow. . . . The depth of the snow in inches as it lies on the ground is frequently wanted in weather predictions, and should be recorded at every observation in addition to the measure of snowfall. . . . "

The above passage is quoted from the 1888 Treatise on Meteorological Apparatus and Methods, by Cleveland Abbe[1].

Weather observers have taken snow observations using similar general guidelines that date back to when nationwide observations began in the 19th Century. But over the years, a variety of subtle changes in how snow observations are taken and recorded have evolved. For example, fundamental differences have existed for many years in how snowfall measurements are taken by primary staffed "First Order" National Weather Service (NWS) weather stations and other aviation weather stations across the U.S. compared to volunteer weather observers in the NWS Cooperative Program. Cooperative observers measure once each day while aviation weather stations report precipitation and snowfall at six-hour intervals. Cooperative observers are instructed to report snow that melts on contact with the ground as "trace" snowfall while the First-Order stations estimate the amount of snowfall assuming no melting. Data from both sources end up in the same published and computerized data bases that are widely distributed and used.

These are just a few examples of observational inconsistencies that are a part of our climate records. A documented history of snow observations, similar to the 1963 "History of Weather Bureau Precipitation Measurements" written by J. H. Hagarty of the U.S. Weather Bureau Office of Climatology[5], has not been written, but would be a valuable reference for climatologists and other users of historic snow data.

Elements of Snow Observations

Standard climate observations of snow consist of daily determinations of three required elements: precipitation, snowfall, and total depth of snow on the ground. In addition to these elements, the measurement of the total water content of all snow on the ground is critical for determining future water supplies and flood potential and for assessing the load that accumulated snow places on man-made structures. The measurement of the water content in a core of fresh snow that has accumulated during the past day has not been a required observation, but those familiar with precipitation gauges know that these core measurements are more accurate than data from regular precipitation gauges under many circumstances and, therefore, should be taken routinely.

The most useful observations of snow also include descriptive remarks. A note of when the snow began or ended may prove helpful in the future. Just the words "blizzard" or "snow melted as it landed" or "wet snow" help climatologists to better interpret and utilize the data. Observers should remember when they observe and record weather conditions that they are documenting local and national weather history.

The following definitions describe the components of a **climatological snow observation.** Other terms related to snow are given in the glossary near the end of this booklet.

Precipitation:
: The accumulated depth of rain or drizzle and also the melted water content of snow or other forms of frozen precipitation, including hail, that have fallen in the past 24 hours or since the previous observation.

Snowfall:
: The depth of new snow that has fallen and accumulated in the past 24 hours or since the last observation.

Snow depth:
: The combined total depth, at the time of observation, of both old and new snow on the ground from a representative location.

Water equivalent of snow on the ground:
: The amount or depth of water obtained by melting a representative core of the snow on the ground.

Water equivalent of fresh snow:
: The amount or depth of water obtained by melting a core of snow that has accumulated, normally on a snowboard, during the past 24 hours or since the previous observation.

Snow core:
: A sample of snow gathered for the purpose of measuring water content. A core may be taken of either just the freshly fallen snow or the combined old and new snow on the ground, obtained by pushing a cylinder down through the snow layer and extracting it. This is sometimes called "cutting a cookie," "cutting a biscuit," or "taking a core" by long-time weather observers.

Whenever possible it is important to measure each of these elements. While certainly related, these elements are each separate variables. It is normally not appropriate to derive one variable from another without measuring it independently. In the past, some weather observers were taught to estimate snowfall by measuring precipitation and then multiplying by ten to get snowfall. Similarly, it has been the policy at some stations to measure the depth of new snow and divide by ten to determine the water-equivalent precipitation. If all snow had the same density and if all snow was captured uniformly by precipitation gauges, such an approximation might be valid. But we have learned that snow changes greatly from storm to storm and from place to place, so such estimates result in measurement inaccuracies. For climatic consistency, it also is best if snow observations are taken at consistent intervals and at the same time each day.

Instruments for Measuring Snow

Many people imagine that all meteorological measurements, including measurements of snowfall, are taken with sophisticated equipment. They are surprised to discover that most snowfall measurements today are taken exactly as they were 100 years ago. The equipment at most weather stations is simple – a human, a simple precipitation gauge, and a measuring stick.

Traditional instruments for manual snow observations

In snowier climates some observers use a snowboard. They are easy to use and maintain. All weather stations in the country that ever receive snow should have and use one. A snowboard is nothing more than a piece of plywood or other light, strong material painted white to minimize radiation and melting effects. Typical dimensions for a snowboard are approximately two-feet by two-feet and approximately one-half inch thick. After each snow observation, the board is cleared of snow and set flush with the existing snow surface.

In the snowiest parts of the country, weather stations also should be equipped with a snow stake. This is a two-inch by two-inch piece of lumber, or comparable material, taller than the deepest snow possible for that location, painted white with inch and foot markers (centimeters, if observations are metric) clearly visible, and permanently installed in the ground in a protected area where snow accumulates uniformly. The observer can read the depth of snow to the nearest inch from some distance away without needing to trudge through the deep snow and risk disturbing the snow surface near the stake.

A sturdy ruler, a good attitude and a basic appreciation for the complexities of snow are the most important instruments needed for taking measurements.

Photo by Daniel B. Glanz

Traditionally, precipitation falling as snow in the United States is measured just as rain would be measured. A standard rain gauge – an eight-inch-diameter collector at all official National Weather Service weather stations, is used to collect the falling precipitation. Recording gauges used at some weather stations are equipped with antifreeze solutions so that the snow is melted as it lands. Non-recording gauges used by most of the nation's official cooperative weather stations require the observer to melt the frozen precipitation that has fallen in the gauge before pouring the melt water into the inner tube for precise measurement. The proper methods for measurement follow later in this text.

There are a variety of other precipitation gauges used by non-National Weather Service weather observers. For the measurement of snow, the important thing is the size of the open cylinder and the exposure of the gauge. Gauges with narrow openings are not good for collecting snow. Also, gauges with permanent funnels do not work since snow will quickly collect and cover the top. If snow continues to fall, it will fall off or blow out without being measured.

Wind shields are used along with precipitation gauges at some locations in the United States to help improve the quality of the gauge measurement of precipitation that falls as snow. For many years, weather observers have used the Alter shield at selected stations to help reduce wind speeds and increase air turbulence at the mouth of precipitation gauges so that a higher fraction of the falling precipitation is collected in the gauge. Studies have shown that other types of shields may work better. In Canada, many weather stations are equipped with a bell-shaped wind shield known as a Nipher shield. Unfortunately, it tends to clog with snow in the heavy, wet snows that characterize many U.S. storms. A special shield for extreme wind environments was developed in Wyoming and is known as the "Wyoming wind shield." It consists of two concentric rings of angled wooden slats with an Alter-shielded gauge in the center. It currently is used at a small number of sites in the Rockies and in Alaska.

Although wind shields have been shown to improve gauge catch, the majority of U.S. precipitation gauges are not shielded due to inconvenience and expense. During the 1990s, an Alter-like shield made of vinyl cloth instead of metal is increasing in popularity. Policies concerning shielding of gauges in the U.S. remain a subject of debate.

A standard precipitation gauge is an essential part of most climatological stations. A wind shield reduces wind speed and increases turbulence near the top of the gauge, which improves the gauge catch efficiency.

Photo from the National Weather Service

A snow stake is a great help in measuring the depth of snow on the ground in snowy climates. This specially designed stake also includes a moveable snowboard for measuring fresh snowfall.

Photo by Ken Dewey

A snowboard significantly improves the accuracy and consistency of snow observations if used properly and located in an area where snow is deposited uniformly.

Photo by Daniel B. Glanz

The traditional instruments for measuring snow consist of a precipitation gauge (outer can, funnel and inner measuring tube), a measuring stick, ruler, warm water to melt snow, and a metal sheet or clipboard to help extract snow cores.

Photo by Daniel B. Glanz

The United States Department of Agriculture, Natural Resources Conservation Service, has a long history of measuring precipitation and water content of snow on the ground. They use special tubes for manual measurements to take core samples of snow on the ground. The core samples are weighed to obtain the measurement of water content of snow on the ground. Since the late 1970s, remote automated measurements of the water content of snow on the ground have been taken using instruments called snow pillows that sense the weight of snow lying on them. These stations transmit their data daily, or more frequently, by radio signals. Precipitation measurements also are taken using tall standpipe gauges charged before each winter with an oil-antifreeze mixture. A sensor in each gauge converts the volume contents of these storage gauges into a calibrated depth. Day-to-day-changes in depth reveal the accumulation of precipitation through the year. These systems have been a great boon to snow monitoring in remote places, but they do require maintenance and care in order to collect reasonable data.

The instruments described above continue to be the primary instruments for measuring snow. Many other instruments and remote sensing technologies can be used to measure selected snow properties. The list that follows is not exhaustive.

Special tubes for extracting snow cores and scales for weighing these samples greatly simplify the measurement of snow water content in snowy areas.

Photo from Natural Resources Conservation Service

A Natural Resources Conservation Service SNOTEL (Snow Telemetry) site in the western United States includes a snow pillow (foreground), a shielded standpipe storage precipitation gauge and radio telemetry equipment.

Photo from Natural Resources Conservation Service

Acoustic or Optical Snow Depth Sensor

This type of instrument transmits an acoustic or optical signal straight down and receives a reflected signal back from the snow surface and provides a determination of the depth of snow on the ground. These systems work well when the snow surface is smooth and uniform and when no precipitation is falling.

Light-Emitting-Diode-Weather Indicators

Falling snow crystals scatter light from narrow light beams. The scatter and scintillation (sparkling or twinkling) of the light beam is related to the size, shape, and fall speed of the precipitation particles. It is possible to determine when snow is falling and to approximate the precipitation and snowfall accumulation rate with this type of instrument when properly calibrated.

Radar (Active Microwave)

The same radar devices used to track rain and thunderstorms can detect falling snow in the winter over distances of tens of miles. Snow does not reflect radar microwaves as efficiently as raindrops and hailstones, so only rough estimates of snowfall location and intensity can be made from this technology at this time. Microwaves transmitted from aircraft and reflected from snow and ice surfaces can also be used to deduce some snow properties.

Visual Satellite Images

Satellites orbiting the earth are able to continuously transmit images back to collection sites on the ground. Since snow is an excellent reflector of sunlight, in the absence of cloudcover, images from space clearly show areas of the earth which are covered with snow. For several climate applications, the spatial extent of snow cover is extremely important. However, there are several limitations to this technology. It is difficult to distinguish deep snow from clouds in areas that are often cloudy. Snow cover also can be difficult to detect properly in extensive coniferous forest regions. Visible imagery is unable to estimate depth or water content of snow on the ground.

Passive Microwave Remote Sensing

When snow covers the ground, some of the microwave energy emitted by the underlying soil is scattered by the snow grains. The degree of scattering

Light-emitting diode weather indicators like this one are a part of the National Weather Service Automated Surface Observing System (ASOS) weather stations.

Photo by Grant Goodge

can be correlated with the amount of snow on the ground. From these relationships, regional algorithms have been developed which indicate the presence of snow and compute estimates of snow water equivalent or snow depth using an assumed density. This technology has limitations. For example, shallow or wet snows may be difficult to detect. The advantage of this method is that microwave energy is not obstructed by cloud layers, so that snow cover can be estimated from space regardless of weather conditions below.

Gamma Radiation Detection

The earth is constantly emitting gamma radiation into the atmosphere from soil surfaces. Snow cover diminishes or attenuates this radiation in proportion to the water content of snow on the ground. The greater the water content, the less the emission of radiation into the atmosphere. Aircraft measurements of gamma radiation can be used to estimate the water content of snow on the ground along the flight path of the plane. This technology has been used for several years to help estimate flood potential from snowmelt in the northern United States.

Photo by T.R. Youngstrom, Telluride Ski Resort

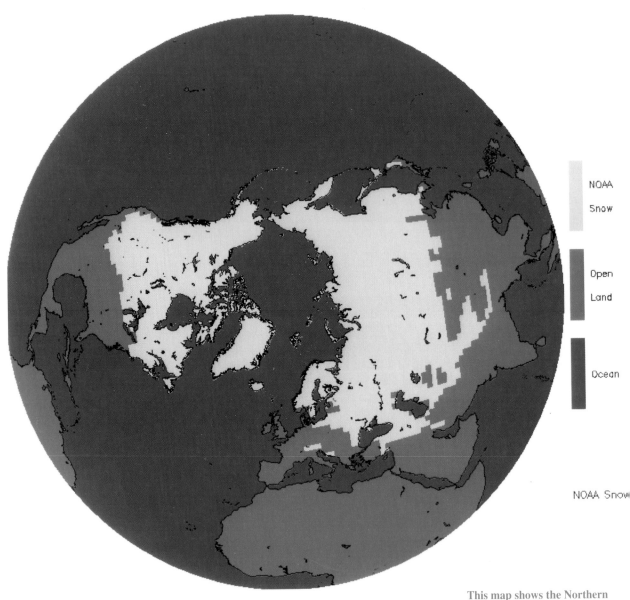

NOAA
Snow

Open
Land

Ocean

NOAA Snow

The monitoring of snow on the ground from satellites provides aerial coverage and spatial detail that is not possible from surface observations. Remote sensing has greatly benefited climatic research. However, manual snow measurements from surface weather stations remain a necessary and valuable part of climate observing.

figure 36a

This map shows the Northern Hemisphere average snow extent for the period January 18-24, 1988, derived from visible-band satellite data (NOAA Weekly Snow Chart).

National Snow and Ice Data Center, Boulder Colorado.

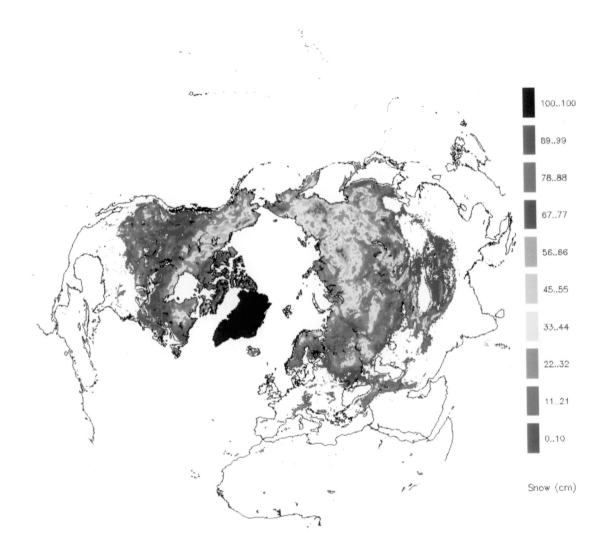

■	100..100
■	89..99
■	78..88
■	67..77
■	56..66
■	45..55
■	33..44
■	22..32
■	11..21
■	0..10

Snow (cm)

**Northern Hemisphere average
snow depth in centimeters for
the period January 18-24, 1988,
derived from passive
microwave satellite data
(DMSP-SSM/I).**

National Snow and Ice Data Center,
Boulder, Colorado

figure 36b

Problems and Challenges in Measuring Snow

To the uninitiated, measuring snow seems simple. All you need to do is push a measuring stick into the snow, then read the number on the stick and write it down, right? Unfortunately, it's not that easy.

Three properties of snow are responsible for most of the difficulties in making accurate and consistent measurements.

1. Snow often melts as it lands or as it lies on the ground, both from warm soil below or from warm air, wind, or sunshine above. As a result, the observer may be in a dilemma. The observer may wonder, "It snowed for three hours, but there is nothing on the ground. Do I report zero or a trace or do I make something up?" (See page 64 for the answer.)

2. Snow settles as it lies on the ground. Depending on the initial density of fresh snowfall and on other coincident weather conditions, the snow may settle rapidly or very gradually. This can have profound effects on observations. Observers who measure more often than once each day may report much more snowfall than a once-daily observer.

3. Snow is easily blown and redistributed. It tends not to land or lie uniformly on the ground, but instead forms deep drifts in some areas while exposed areas may be blown completely clear. A related problem is precipitation gauge catch. Even at relatively low wind speeds, snow is easily deflected around the top of precipitation gauges. Most winter precipitation measurements using standard precipitation gauges underestimate the precipitation that actually fell and sometimes by large amounts (see Figure 41.).

Because of these three factors, the location where observations are taken, the time of day when observations are taken, and the frequency with which observations are taken (once daily for most cooperative stations, every six hours for National Weather Service First Order stations and even more often than that for some over-zealous snow lovers) all affect the data. Traditionally, National Weather Service stations at major airports around the country use a midnight-to-midnight observing schedule for reporting daily snowfall. However, only a fraction of the several thousand official cooperative stations report at midnight. The majority of cooperative observers take their daily readings in the morning between 6:00 a.m. and 9:00 a.m. A substantial number also take their observations in the late afternoon or early evening. Historic snowfall data in the U.S. would be more consistent if all observers took their daily measurements at approximately the same time each day.

Caution: Official NWS cooperative observers should not change their scheduled observation time without NWS approval since the time of observation affects other climatic data including average temperatures.

"He used to love snow as much as me — and then he became a weather observer."

figure 37

Average annual snowfall at Fort Collins, Colorado based on maximum accumulation of snow during the day from six-hourly observations (A) compared to snow totals derived from daily measurements of the 24-hour increases in total depth of snow on the ground from 7:00 a.m. to 7:00 a.m. (B) and from 7:00 p.m. to 7:00 p.m. (C). These results are based on ten years of observation, 1985-94.

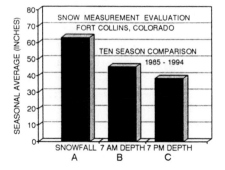

In order to accurately measure and record snowfall, an observer must know what he or she is supposed to measure and must have a representative location to take those measurements. Do we include snow that melted even though it disappeared before we had a chance to measure it? For snow that is settling over time, do we report its current depth or its maximum depth prior to the observation? For snow that has drifted, do we use our measurements from our normal point of observation or do we average over a much larger area? These questions all are a part of the snow measurement challenge.

Impact of Inconsistent Data Collection

The difficulties described above may have direct impacts on the snow data that we collect and on the data used in past, present and future climate studies. Some of the effects are subtle. Weather stations that take their daily observations in the morning tend to measure more snow than those that measure in the late afternoon. More rapid daytime melting and settling cause these differences.

Some observers determine daily snowfall by simply noting the change in total snow depth from one day to the next. These observers routinely report less snow than those that measure new snow on a snowboard or those who measure more frequently than once daily.

Over periods of many years, weather stations may be moved. Changing the location and exposure of the weather station can affect snowfall reports. A move from a protected area to a more open windswept location often will result in less measured snowfall due to wind effects. In coastal or mountainous areas, station moves of very short distances may result in large changes in observed snowfall that are real.

Changing from one weather observer to another can make a surprisingly big difference. It seems that some observers who enjoy snow and watch it constantly tend to report more snow than observers who go outside only once a day to take their measurements. National Weather Service offices for many years have monitored weather conditions continuously and taken measurements at six-hour intervals. These sites typically report more snowfall than volunteer stations that measure only once each day.

Differences in snowfall totals between the First-Order National Weather Service Station at Wilmington, Delaware, and the nearby cooperative weather station at Porter Reservoir for nearly 400 snow events. The First-Order station, which takes measurements at six-hour intervals, consistently reports more snowfall than the cooperative station that measures just once each day. These data were provided by Daniel Leathers, Delaware State Climatologist.

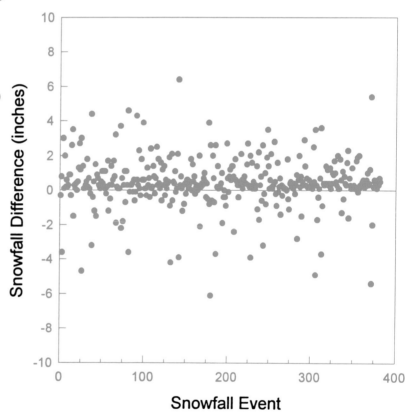

One way to spot observational inconsistencies is by looking at the ratio of reported precipitation to reported snowfall. This ratio can be called the apparent density of freshly fallen snow because it uses snow water content derived from gauge measurements instead of snow cores from snow on the ground. Apparent densities are normally lower than actual densities since gauges tend not to catch all the precipitation that falls.

If apparent densities averaged over a month, a winter season, or over several years are greatly different between nearby stations, or if a longterm time series at a single site suddenly changes significantly, it suggests there are problems with how one or both of the observations have been taken.

Collecting snow data is sufficiently difficult that few longterm records dating back more than 80 to 100 years are available for studying patterns and changes in snow in the United States. There are only a few hundred weather stations in the United States with snow data that are complete back to 1900. Most of these sites have inconsistencies due to station moves, time or frequency of observation changes, exposure changes, or changes in how snow was measured and recorded.

figure 39

Major changes in observed snow density may signify observational problems. A change in observer and station location in Leadville, Colorado, in 1985 was accompanied by a dramatic decrease in the apparent densities of freshly fallen snow. It has since been learned that the new site experienced stronger winds and more drifting than the previous more protected site. Snow cores were not taken at this station to improve precipitation estimates. As a result, gauge precipitation has been considerably lower than what actually fell. Snowfall estimates have been exaggerated due to drifting.

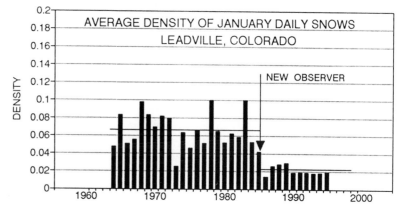

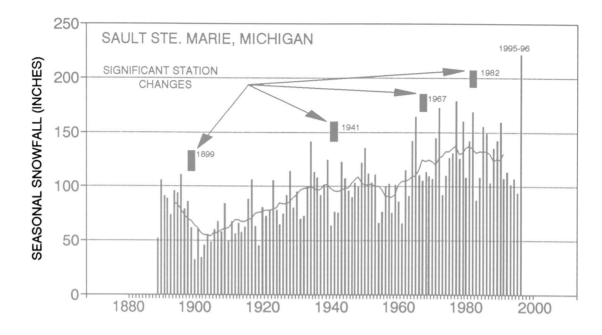

figure 40 Historic seasonal snowfall totals at Sault Ste. Marie, Michigan. A combination of station moves, changes in exposure, and changes in observational procedures (some of which have not been documented) are superimposed on what appears to be a very significant longterm increase in snowfall. Longterm snowfall time series for other parts of the country are plagued by similar problems. Improving the consistency in observations would make it easier to confidently identify important climate variations and trends.

The result is, that despite more than 100 years of nationwide climate monitoring in the United States, we don't have the quality and quantity of snow data available for accurate comparisons or longterm studies of snowfall, snow depth, snow cover, and water content. The effects of inconsistent observing and reporting are profound. The differences are not just a few percent. Instead, winter precipitation and snowfall totals may differ by 25 percent or more depending on station exposure and observing procedures. In some cases, differences exceed 100 percent.

We need to do better. We can improve snow data in the United States if a concerted effort is made on a national basis to standardize observing procedures and to train observers in proper observing methods.

Procedures for Measuring Snow

The following procedures were developed based on the expertise of many climatologists, weather observers, snow specialists, and data users. These are based on the wealth of oft-forgotten knowledge and experience from several generations of snow observers. These suggested procedures are very similar to the standard instructions for measuring snow that have been in use within the National Weather Service for many years, but with a few important additions and changes. These we feel will significantly improve snow data collected in the United States.

Basic Preparations

Preparation is the key to accurate and consistent snow observations. Anyone can take good snow observations with the help of a little enthusiasm, some basic instruction, simple equipment, an appropriate and representative location to take observations, and a little knowledge about how snow behaves.

When your weather station is first established:

Invest time in carefully determining the best location for installing the precipitation gauge (see page 65). Finding a good site at the very beginning will improve and simplify observations from then on.

In order to select the best location(s) for measuring snowfall and depth, carefully consider drifting snow and wind patterns and sunshine and shading patterns.

Establish and maintain a fixed time of day for completing the daily weather observation. Midnight has been the standard for NWS First-Order stations and is very convenient for record keeping. It would be ideal if all stations utilized the same daily observation time nationwide, but for volunteer stations this is not a requirement. For volunteer stations, it is most important to select an observation time that is compatible with the daily schedule of the observer and acceptable to the NWS and others who may use the data. Early morning or evening observation times are most common among NWS Cooperative stations.

Each year before the first snow comes:

Review the instructions for measuring snow. It is easy to forget what needs to be measured, especially in those parts of the country where snow falls infrequently.

Check your equipment. Make sure everything is in good condition. Put a new coat of white paint on your snowboard if it needs it. Make sure your precipitation gauge has no leaks. If possible, try to have an extra outer cylinder for the precipitation gauge on hand. If you are in deep snow country, make sure your snowstake is legible and securely installed in the vertical position. Many snowstakes can be found lying on the ground at the end of the summer, which doesn't do much good when the winter snows arrive.

Growth of vegetation and the construction or removal of buildings can change station exposure from one year to the next. Check to make sure that your instrument exposure is still satisfactory. Make improvements if necessary, but always document any changes to the station.

Before the first snow arrives, put out your snowboard and remove the funnel and inner measuring tube from your standard precipitation gauge.

Find a friend or neighbor who shares your interest and can help you out when you need a second opinion or when you are away from your station.

Taking Observations

Here is a simplified list of steps for taking a complete daily climatological observation of precipitation and snow. A detailed discussion of each component of the observation then follows. The lists of special concerns and precautions that accompany each item are very important. Keep in mind that weather conditions at the time of observation may make it impossible or inappropriate to follow these steps in this precise order. Use common sense and good judgement.

Some instructions include the words "measure and determine." In the observation of snow, a single *measurement* at the time of observation may not be sufficient to accurately assess snowfall or depth. It may be necessary to "determine" these quantities by using additional information such as the average of many measurements or measurements taken at previous times during the day.

Steps in completing a daily climatological precipitation and snowfall observation:

1. Measure and determine the amount of snowfall for the previous 24 hours.

2. Measure and determine the total depth of snow on the ground (the sum of both the old snow, plus any new snow that has accumulated) at the scheduled time of observation.

3. Measure the amount of precipitation that has accumulated in the precipitation gauge during the previous 24 hours.

4. Measure the water content of the new snow that has accumulated on the ground during the previous 24 hours.

5. Measure the total water content of both old and new snow on the ground.

6. Record all observations, using the appropriate units, on a climatological observation form. Make certain that the form contains all necessary information to accurately identify the correct weather station, day, month, and year. Double check all entries for accuracy and consistency. Include descriptive remarks.

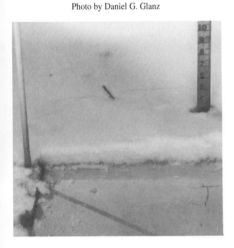

Measure daily snowfall by determining the maximum accumulation of new snow during the past 24 hours using a snowboard, if possible.

Photo by Daniel G. Glanz

7. If previously arranged, transmit observation by phone or computer to the National Weather Service, local media and/or other interested sources.

Snowfall

Snowfall is measured daily at the specified time of observation, or soon after snowfall ends, using a sturdy ruler and a snow board. The observer must attempt to find a protected and representative location for observing snowfall where the effects of wind are minimal. The goal in measuring and reporting daily snowfall is to observe the *maximum accumulated depth of new snow since the previous day* before melting and settling reduced the depth.

A snowboard laid on the ground or on the surface of old snow allows for a more precise determination of snowfall. The snowboard must be flush with the surface or measurements will not be accurate. Hold the measuring stick vertical and push it gently into the snow until it reaches the surface of the snowboard or the bottom of the layer of new snow. Read this depth to the nearest 0.1 inches. This depth is the daily snowfall that should be recorded unless blowing, drifting, melting, or settling have significantly altered the accumulation. If these have occurred, and they often do, the observer must use good judgement and include additional information to make a better determination of snowfall. See items one through four below and "Dealing With Adversity," pages 73-76, for determining snowfall under such conditions.

Special concerns and precautions:

1. It is essential to measure snowfall in representative locations where the effects of blowing and drifting are minimized. Finding a good location where snow accumulates uniformly simplifies all other aspects of the observation and reduces the opportunities for error. In open areas where windblown snow cannot be avoided, an average of several measurements may often be necessary, not including the largest drifts.

2. Snow often melts as it lands on the ground. If snow continually melts as it lands, and no accumulation is ever noted even on grassy surfaces, snowfall should be recorded as a "trace," and a remark should be entered, "Snow melted as it landed."

3. If snow partially melts as it lands, but some accumulation takes place, the snowfall for the day should be recorded as the *greatest accumulation of new snow observed at any time during the day.* When snow accumulates, melts, and accumulates again, the snowfall is the sum of each accumulation before melting.

4. In addition to melting, snow settles and compacts as it lies on the ground. The preferred record of daily snowfall should be *the maximum accumulation of new snow observed at any time during the day.* Since volunteer observers are not always available to watch snow accumulation at all times of the day and night, observers' best judgment must be used based on a measurement of snowfall at the scheduled time of observation along with knowledge of what took place during the day. If the weather observer is not present to witness the snow accumulation, input should be obtained from other people who were near the station during the snow event.

5. Never use the sum of frequent snow observations of a cleared surface to determine daily snowfall. This will invariably inflate the apparent snowfall and provide unreasonably low ratios of snowfall to water content.

6. If a snowboard is not available, wooden decks or platforms and grassy surfaces are usually good alternatives. Beware that snow may perch on top of a grassy surface. A measuring stick will penetrate through both snow and the airspace among the blades of grass resulting in an overestimate of snow.

7. Summer hail is not reported as snowfall. Winter ice pellets and sleet do count as snowfall.

8. In areas prone to heavy snow, the snowboard should be marked clearly with a flag or have an attached stake or pipe that is taller than the maximum expected snowfall. This way, you should never loose the snowboard in deep snow, and it is easier to handle while taking snow cores.

Depth of Snow on the Ground

The total depth of snow on the ground at the time of observation is normally the easiest of the snow measurements, except when significant drifting has occurred. Snow depth is measured by reading the depth of snow at the permanently-mounted snow stake or by taking the average of several depth readings at or near the normal point of observation. Depth of snow on the ground is measured and reported to the nearest whole inch.

To determine the precipitation from snow that has landed in the precipitation gauge, the contents of the gauge must be melted using a measured amount of warm water.

Photos by Daniel B. Glanz

1. The key factor is finding a location for the measurement so that the snow depth that is reported is the average depth of snow for unshaded, level areas, not disturbed by human activities, in the vicinity (within several hundred yards) of the weather station.

2. During snowmelt, observers often face the situation where snow melts quickly from south-facing areas, but remains deep in shaded or north-facing areas. An observer should use good judgement to visually average snow depths in the area surrounding the weather station. For example, if half the ground is bare and half the ground is covered with four inches of snow, an appropriate snow depth report would be two inches. When less than 50 percent of the ground area is covered by snow, then snow depth should be recorded as a "trace."

3. When new snow falls atop a layer of partially melted uneven old snow, it can be very difficult to find an appropriate point or area to measure snow depth. Many measurements may be needed to obtain a valid average, or observer judgement may be required to estimate a value. Stay away from paved areas and areas with lots of human or animal footprints in the snow.

4. The depth of hail on the ground at the time of observation can and should be recorded in the same way as snow depth. Make a remark that identifies the accumulation on the ground as hail.

Precipitation

Precipitation is measured daily at the specified time of observation using an appropriate precipitation gauge. At most locations in the United States, a standard gauge made of copper or other material with a round opening eight-inches in diameter is used. In climates where snow falls, the inner tube and funnel should be removed during the snow season. When snow or other frozen precipitation has fallen, the contents of the gauge are melted and poured into the inner tube for measurement. Precipitation should be read and recorded to the nearest 0.01 inches using the measuring stick provided with the gauge.

For accurate measurement, it is imperative that the precipitation gauge have a good exposure not too close to buildings or trees, but protected from strong winds. If possible, the gauge should be at least twice the distance away from objects as the height of those objects. For example, if a nearby barn is 25-feet high, the gauge should be at least 50-feet away. Forest clearings usually are good locations. In more urban areas, back yards in residential neighborhoods with bushes and shrubs, but few tall trees are acceptable locations. If natural shielding from the wind is not possible, then wind shields

After melting snow in the gauge, the contents must be carefully poured into the inner cylinder for measurement. The amount of warm water that was added to melt the snow must be subtracted from this total. The remainder is the precipitation total for the day from the precipitation gauge.

Photos by Daniel B. Glanz

are recommended. Rooftop exposures are never recommended, even if a wind shield is used.

Special concerns and precautions:

1.   Melting the snow in the gauge before measuring it can be tricky. Do not set the gauge on a stove or hot plate to melt the snow since it could damage the gauge and might cause some of the water to evaporate before it is measured. An effective method for melting the snow is to add warm water to the gauge. When using this method, observers must carefully measure the amount of water added to melt the snow and not include that amount in the precipitation measurement. It is slower, but equally effective to set the gauge in a bucket of warm water. Make sure you dry off the outside of the gauge with a towel before pouring the melted contents into the measuring tube.

2.   It is easy to spill water when pouring from the large outer gauge into the inner measuring tube. Be very careful. Secure the inner tube so it cannot fall over, and use the funnel to make sure contents are not spilled.

3.   National Weather Service measuring sticks look like ordinary rulers, but they are not scaled like ordinary rulers. They are calibrated for use in the inner measuring tube only. The area of the open end of the inner tube is only one-tenth the area of the large outer cylinder, so when pouring from the outer can to the inner tube the depth is increased ten times. The special calibrated ruler takes this into account. Do not let this confuse you. **Errors in decimal point placement must be avoided at all costs.** They are a huge problem in our nation's historical precipitation data sets. If observers make this type of mistake, they should be able to catch and correct the problem by noting the unrealistic ratios of precipitation to snowfall that result.

4.   Do not measure the melted precipitation directly in the large outer cylinder. The melt water first must be poured into the inner cylinder for proper measurement. (Exception: There are a few weather stations equipped with scales that convert the contents of the eight-inch gauge directly into hundredths of an inch of water equivalent precipitation without needing to melt it.)

5.   Precipitation gauges do not collect all precipitation that falls. The gauge-catch efficiency decreases as wind speeds increase and as the density of freshly falling snow decreases. The stronger the wind and drier the snow, the poorer the gauge catch. Even light winds of only three miles per hour at the top of the precipitation gauge are enough to reduce gauge catch by 30 percent. There are examples in the northern plains and in wind-scoured areas of Wyoming, Montana, and Colorado where observers have found only trace amounts of snow inside the eight-inch gauge following blizzards that produced snowdrifts several feet high. That is why it is so important to find a well-protected site for

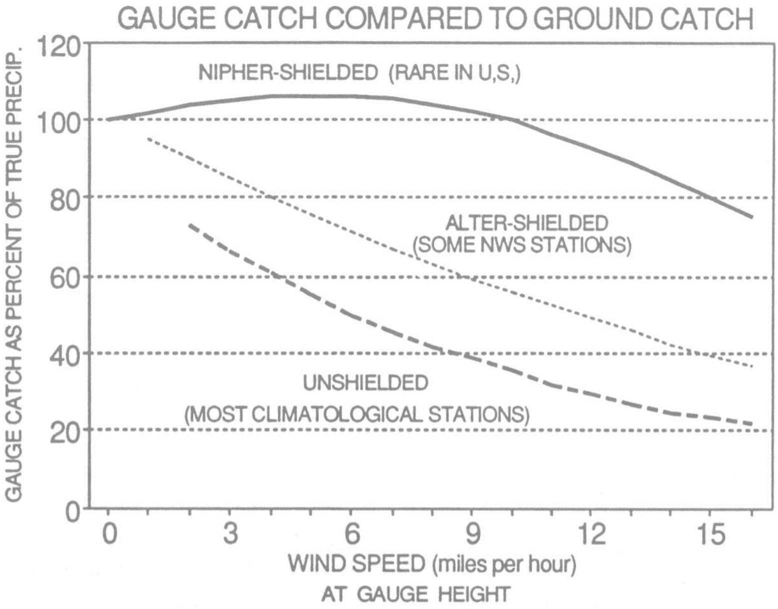

GAUGE CATCH COMPARED TO GROUND CATCH

NIPHER-SHIELDED (RARE IN U,S,)

ALTER-SHIELDED
(SOME NWS STATIONS)

UNSHIELDED
(MOST CLIMATOLOGICAL STATIONS)

GAUGE CATCH AS PERCENT OF TRUE PRECIP.

WIND SPEED (miles per hour)
AT GAUGE HEIGHT

Precipitation gauges collect less precipitation than what actually falls when winds are strong. If the gauge is unshielded, like most cooperative stations in the U.S., precipitation gauge measurements become almost meaningless when winds near the gauge are strong (after Goodison, 1978)[2].

figure 41

installing the gauge and why it also is recommended to measure the water content in core samples of freshly fallen snow.

6. Wet snows that stick to the rim of the gauge decrease the size of the opening, which reduces the amount of snow collected. If an observer notes snow accumulation beginning on the rim of the gauge, this snow should be removed. An objective way to determine how much snow on the rim should fall into the gauge and how much should fall outside is to press a flat board or book, at least ten-inches across, straight down on top of the gauge.

7. In parts of the country that are prone to very heavy, wet snows, such as the Cascade Mountains and Sierra Nevada, the tops of precipitation gauges can totally cap over with snow. After the gauge is capped with snow, additional snowfall simply blows or slides off the top and is not measured and causes large measurement errors. A similar problem can occur anywhere in the country if observers are caught off guard by heavy snow and still have the funnel installed on the top of the gauge. After four to six inches of snow accumulates in and on the funnel, additional snowfall may be lost and its water content not measured.

8. Gauges can develop leaks and most eventually will. Observers must check both the eight-inch outer can and the inner cylinder for leaks on a regular basis. If water is allowed to freeze solid inside the gauge, the chances of developing a leak increase.

SNOW MEASUREMENT GUIDE

OBSERVERS WITH NON-RECORDING GAGES RECORD THREE MEASUREMENTS WHEN IT SNOWS

(1.) WATER IN THE SNOW

Record in this column to inches and hundredths.

Melt contents of gage and measure like rain. If high winds have blown snow out of the gage, the outer container is used to obtain a substitute sample from the snow on the ground where the depth represents the amount that fell since yesterday's observation.

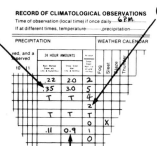

RECORD OF CLIMATOLOGICAL OBSERVATIONS
Time of observation (local time) if once daily——— 6 P.M.
If at different times, temperature————,precipitation————

PRECIPITATION			WEATHER CALENDAR			
ved, and a served	24 HOUR AMOUNTS		Fog	Sleet	Glaze	Thunder
10 11	Rain Melted Snow etc (ins. & hundr'ths)	Snow Sleet Hail (ins. & tenths)	Snow Sleet Hail on grnd (inches)			
	.22	2.0	2			
	.35	3.0	5			
	T	T	4			
			2			
	T	T	T			
			0	X		
	.11	0.9	1			
			0			

(3.) DEPTH OF SNOW ON THE GROUND AT OBSERVATION TIME

Record in this column to nearest inch—if less than ½ inch, record "T".

Any time there is snow on the ground at observation time record average depth on ground at observation time. Include old snow as well as newly fallen snow.

(2.) SNOWFALL SINCE YESTERDAY'S OBSERVATIONS

Record in this column to the nearest 0.1 inch.

Find some place where the freshly fallen snow is least drifted and is about average depth for the locality. Measure the depth of the snow which fell since yesterday's observation. Report an estimate if the snow melted before observation time.

When significant amounts of new snowfall have occurred round off to the nearest inch and record as, for example, 2.0 and 3.0. (Record as 2.0 not 2, 3.0 not 3).

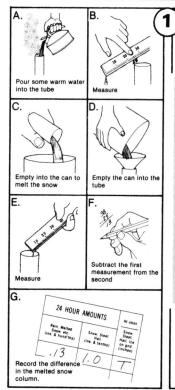

A. Pour some warm water into the tube

B. Measure

C. Empty into the can to melt the snow

D. Empty the can into the tube

E. Measure

F. Subtract the first measurement from the second

G.

24 HOUR AMOUNTS		At obsn
Rain, Melted Snow, etc. (ins. & hundr'ths)	Snow Sleet Hail (ins. & tenths)	Snow Sleet Hail, Ice on grnd (inches)
.13	1.0	T

Record the difference in the melted snow column.

(1)

At the beginning of the snowfall season only the 8-inch gage can is exposed to catch the snow. The funnel and measuring tube are removed at the beginning of the snowfall season. The measuring tube is used to measure the water from the melted snow.

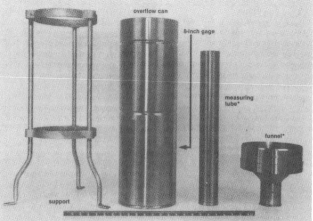

overflow can

8-inch gage

measuring tube*

funnel*

*removed during winter months

snow won't fall in representative quantity into the gage if the funnel and measuring tube are not removed.

(See reverse side for steps 2 and 3)

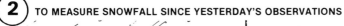

2 TO MEASURE SNOWFALL SINCE YESTERDAY'S OBSERVATIONS

1. If the snow melts as it falls, enter a trace for snowfall.

2. Measure each new snow. Use good judgment in selecting spots where the snow is least affected by drifting.

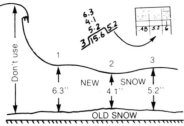

3. When possible, take several measurements where the snow is least affected by drifting (don't include deep drifts) and average.

4. If the snow has blown out of the can or the "catch" is not good, cut a "biscuit" with the can where the snow is near the average and melt the biscuit for the water equivalent.

3 SNOW DEPTH

Entry in this column is the measurement to the nearest whole inch of all snow, sleet, ice and hail remaining on the ground at your regular observation every 24 hours.

24 hr amounts		At obsn
Rain Melted Snow etc (ins & hund'ths)	Snow, Sleet Hail (ins & tenths)	Snow, Sleet, Hail, Ice on gnd (inches)
.32	T	0

Rain and snow mixed; snow melted as it fell.

24 hr amounts		At obsn
Rain Melted Snow etc (ins & hund'ths)	Snow, Sleet Hail (ins & tenths)	Snow, Sleet, Hail, Ice on gnd (inches)
		T
.16	2.0	0

2.0 inches of new snow fell, containing .16 water--snow melted before time of observation.

24 hr amounts		At obsn
Rain Melted Snow etc (ins & hund'ths)	Snow, Sleet Hail (ins & tenths)	Snow, Sleet, Hail, Ice on gnd (inches)
		T
		T
.27	1.8	2
		1

1.8 inches snow and ice pellets containing .27 water. 2 inches on ground at observation time.

This useful guide has been provided to National Weather Service Cooperative observers for many years. In simple form, it provided many of the same instructions described in this book. The primary differences are that we recommend the use of snowboards, we define snowfall as the maximum accumulation of new snow, before melting and settling reduce the depth, and we encourage routine measurements of water content of new snow from snow cores and total water content of all old and new snow on the ground.

9. Recording rain gauges are used at some U.S. stations. For best results, an environmentally appropriate light-weight oil and antifreeze should be added to the collector reservoir inside the gauge. The oil produces a film that limits evaporation and the antifreeze keeps the entire contents of the gauge unfrozen for best results.

When snow sticks to the rim of a precipitation gauge, it normally decreases the size of the opening and decreases future gauge catch. Under some circumstances, however, snow may build up on the rim, increasing the rim dimension and then all fall into the gauge, increasing gauge catch. To improve measurements, care must be taken to avoid snow buildup on the rim.

Photo by Grant Goodge

Water Content of Freshly Fallen Snow

This measurement, although not traditionally a requirement from most United States weather stations, is very educational and is essential for many applications. It is an independent measure of the precipitation that has fallen in the past 24 hours that is free of the observing problems associated with precipitation-gauge undercatch. The measurement requires taking a core of fresh snow from a snowboard set on the surface of the old snow layer or on the ground if no old snow is present.

Take a snow core from the snowboard using the eight-inch-diameter outer cylinder of the precipitation gauge. Push the cylinder straight down until it is firmly in contact with the snowboard. Push aside snow outside the ring of the cylinder. Slide a thin, flat sheet of metal beneath the cylinder, scraping up any ice or snow stuck to the surface of the snowboard. Then, holding the metal sheet tightly against the opening of the cylinder so that nothing spills, turn the cylinder right side up again. Remove the metal sheet, making sure that ice and snow which may have stuck to the sheet within the ring of the cylinder fall into the cylinder. If no metal sheet is available, invert the cylinder and snowboard together. Make sure that ice and snow which may have stuck to the board within the ring of the cylinder fall into the cylinder. Proceed to melt and measure the snow in the same manner as a precipitation measurement is taken.

Special concerns and precautions:

1. Under windy conditions or when snow is melting on the snowboard, but not on the ground, the snow accumulation on the snowboard may not be representative of the actual daily snowfall. It may then be necessary to find a more representative location and take a core from the ground without the aid of a snowboard. One must be careful not to accidentally sample the underlying layer of old snow.

2. All the challenges and difficulties associated with measuring precipitation in a gauge still are present. The contents must be properly melted and carefully poured into the inner cylinder without spilling. Any warm water added to the sample to melt the snow must be subtracted from the final measurement in order to give a proper reading.

3. If significant melting has occurred from the time the snow fell to when the core measurement is taken, water that fell as snow may not be in the core sample. This condition should be noted in observer remarks.

Water Content of Snow on the Ground

This measurement requires taking a core of both old and new snow on the ground and melting and measuring the total water content. A constant challenge in determining the water content of snow on the ground is in obtaining a representative measurement in terms of both the snow depth and water content. Normally, the snow core should be taken using the eight-inch cylinder pushed straight down to the level ground surface below. Care must be taken to get all the snow and ice into the sample. Typically, and especially with older snow, the layer closest to the ground contains the most water so this layer must be sampled. It may be useful to clear the area immediately around the cylinder before removing it and turning it over. This will enable the observer to collect any ice or snow that falls out of the cylinder. It is very helpful to have a firm, thin sheet of metal to slide between the surface of the ground and the rim of the cylinder to cleanly gather the entire sample. As you lift the cylinder, tilt it at a 45-degree-angle or more to keep the snow from sliding back out and press the metal sheet firmly over the opening of the cylinder until it is right-side up again. The cylinder contents then may be melted and measured as described previously.

Whenever possible, measure the water content of freshly fallen snow by taking a core from your snowboard. Be careful to obtain a complete core cut cleanly to the surface of the snowboard. Melt and measure the depth of water in the same way as measuring gauge precipitation.

Photo by Daniel B. Glanz

For snow that is deeper than 12 to 18 inches, but less than 24 inches, it is much easier to take a core sample using the small-diameter inner tube instead of the large outside cylinder. The core is easier to extract and is quicker to melt. However, if the inner tube is used, it is only possible to read the water-equivalent depth to the nearest 0.10 inches. The reading on the calibrated measuring stick must be multiplied by ten since no magnification of the sample has occurred.

In those parts of the country where snow depths exceed two feet, it is very awkward and time consuming to use the standard rain gauge for taking snow cores. Special tubes to core the snow and scales to convert the weight of snow cores into equivalent water depth greatly simplify this measurement (see photo on page 51).

Special Concerns and Precautions:

1. A measurement of total water content of snow on the ground is extremely valuable for hydrologic, climatic, and engineering applications, so it is critical that a representative measurement be taken. Without care, it is possible to grossly undermeasure or overmeasure the water content. Make sure that snow does not fall out of the gauge as you remove the core and also make sure you don't inadvertently include extra snow.

2. It is helpful to make an independent estimate of water content based on snow depth and likely density. For example, 20 inches of fresh snow on the ground may have only one to two inches of water content, but 20 inches of aged, compacted snow could contain as much as 3 to 10 inches of water.

3. When no melting has occurred and no fresh snow has fallen, water content changes very little from day to day. Always compare your readings from day to day to check consistency. Any large variation is usually the result of unrepresentative snow sampling.

4. Be sure to cut a clean core all the way to the ground surface. The snow closest to the ground may contain the most water.

5. Care in melting and measuring the snow is required. When the water content is large, greater than one inch, it may take a long time to melt the sample or it may require a significant volume of warm water.

6. Standard precipitation gauges cannot be used for coring snow and measuring water content in areas where total snow depth exceeds approximately 24 inches. Special tubes for coring snow should be used in deep snow environments according to their specific instructions. It is much easier to simply weigh the sample than to melt the sample and measure water depth.

Taking a snow core of both old and new snow using the eight-inch diameter outer gauge cylinder is no easy undertaking when the snow gets deep. A narrower tube and scales to weigh it work well for observers in snowy climates.

Photo by Daniel B. Glanz

Dealing with Adversity

Windblown snow may be the observer's greatest challenge.

Photo by Walter Johnson

There are many problems associated with the collection of snow data in the United States. The fact that observers and observing networks are inconsistent in handling the challenges presented by wind, melting, and settling has resulted in data that may not be accurate for many applications. Just because measuring snow may be difficult is no excuse for accepting inconsistent observations. Thousands of engineers, consultants, managers, resource planners, hydrologists, climatologists, biologists, transportation specialists, and many others rely on and expect accurate data for operations, design, planning, management, and research every year. Standardizing snow observations based on the known properties of snow is feasible if 1) priority is given to finding representative sites for collecting snow data and 2) education about snow and training on how to handle challenging situations are provided.

It is impossible in a short book to describe every difficult situation that will be encountered while observing snow, but here are two stories that present examples of some common situations. These stories show that by using normal procedures accompanied by logical thinking, reasonable and consistent results can be achieved even under adverse weather conditions.

The Big Blizzard

A veil of gray clouds gradually thickened during the day. The winds that had been light and nearly unnoticeable began blowing gently, but steadily from the southeast. As the day progressed, the wind direction shifted gradually to the northeast. Just before sunset it began to snow lightly. Within two hours the ground was white and the sky was obscured by smaller more numerous flakes.

After a hearty supper, the volunteer weather observer for the local community jotted down the time the snow began. Although his official observation time for his daily report was 8:00 a.m., he went outside about 9:00 p.m. and noted that there already was a uniform four inches of fresh snow on the ground. By then it was snowing so hard he could see only the street lights within a half-mile of his house. Walking through the snow, it had a solid feel to it. It did not poof up as he walked, but lay firmly on

the ground. He went to bed early knowing the next day could be an adventure. As he tried to fall asleep the house seemed drafty. The winds increased and the snow brushed against his window with the sound like someone sweeping with a coarse broom. The blizzard was just beginning.

He got up much earlier than usual to see what the storm had accomplished. Indeed, all he could see was solid white as he first looked outside. The temperature had dropped to 17°F.

When the observer fixed his eyes through the blowing snow, he could just make out his old National Weather Service copper rain gauge in the backyard. He bundled up, grabbed his measuring stick, and headed outdoors. Near the back door the snow was nearly two-feet-deep, but soft and fluffy. A little farther out was a wooden picnic table where he often took his snowfall measurements. He once had a snowboard, but it had rotted from neglect. Too bad. The snow under the table was deep, but on top, where he usually measured, the snow had been blown clear. Even the four inches that had been there last night had blown away.

He continued out to his National Weather Service precipitation gauge. Walking became easier. Here only an inch or two of crusty snow covered the grass. Usually, by the time he reached his gauge, he had decided how much snowfall to report, but today he had measured anywhere from nothing to 24 inches on his short walk. What was he going to do? He decided to wait with that decision and gather more information. Perhaps the contents of his gauge would help him.

Inside the eight-inch-diameter receptacle he got another surprise. Fortunately, he had remembered to remove the funnel and inner cylinder from the gauge in the fall when cold weather was setting in. Had he not removed the funnel, most of the snow would have blown away without collecting in the can. He hoped to estimate the snowfall by sticking his ruler down into the gauge, but soon saw that wasn't going to work. Despite the heavy snow overnight, there was surprisingly little in the gauge and it was all jammed up on one side.

He quickly picked up the gauge and hurried inside with it to melt the snow and measure its water content. Some weather stations have two eight-inch-diameter gauges to aid in winter snow observations. But he had only one, so he would need to melt and measure the contents as soon as possible to get the gauge back into the yard so he wouldn't miss more of the snow.

Once inside the house, he poured warm water into the measuring tube, measured 0.51 inches with the calibrated measuring stick and wrote down that amount so he wouldn't forget. He then poured the warm water into the eight-inch can. It was just barely enough to melt the snow. Next, he carefully poured the entire volume back into the measuring tube. He ended up with 0.88 inches on the measuring stick. Subtracting the 0.51 inches of warm water that he had

added left him with a total of 0.37 inches. This was his measurement of precipitation for the past 24 hours which was to be recorded on his observation form. He was confident he had done everything properly, but it just didn't seem right. How could it snow heavily for 12 hours and add up to only 0.37 inches?

He took the emptied eight-inch gauge receptacle back outside and put it in its mount. He scanned the yard more carefully. The wind and snow were diminishing a bit which gave him a better view. He took the measuring stick and took a measurement every few steps. Readings varied from as deep as 26 inches to only 1.2 inches in some windblown areas. For curiosity he dug down through the snow with his gloves and noted that some snow close to the ground was melting, even though the temperature was now far below freezing. He averaged together his measurements and came up with something between nine and ten inches. That seemed reasonable considering how long it had been snowing hard, but that didn't match up well with his 0.37-inch-precipitation measurement. That would be extremely low-density snow, just 0.04, and probably impossible with such strong winds and considering some snow near the ground already melted.

Out of curiosity, he walked beside the house and around to the front. There he noticed the snow was not as deep. There had been more accumulation behind the house since it was more protected from the wind. An overall average snow depth of between eight and nine inches seemed more reasonable. The actual snowfall amount probably was a bit greater due to the melting and settling, but it was impossible for the observer to be certain of that.

The observer now was regaining his normal confidence. He headed back to the house and brought out the measuring cylinder of his rain gauge. He went beside the house and found an area where the snow depth was precisely eight inches and the snow was not severely windpacked. There he pushed the measuring tube straight down into the ground and pulled out a

PRECIPITATION						WEATHER (Calendar Day)							RIVER STAGE				
24-HR AMOUNTS			At Ob.	Draw a straight line (———) through hours precipitation was observed, and a waved line (∿∿) through hours precipitation probably occurred unobserved.		Mark 'X' for all types occurring each day.								GAGE READING AT	CONDITION	TENDENCY	REMARKS (Special observations, etc.)
Rain, melted snow, etc. (Ins. and hundredths)	Snow, ice pellets (Ins. and tenths)	Snow, ice pellets on the ground		AM — NOON — PM		Fog	Ice Pellets	Glaze	Thunder	Hail	Damaging Winds	Time of observation if different from above		A.M.			
0.37	8.0	7															blizzard conditions 2 ft. drifts, gage catch low, snow core approx 0.65"

cylinder of snow (snow core). He took this inside and melted it. It would have been more accurate to have cut a snow core with the eight-inch can and poured the melted contents into the inner cylinder to measure precisely, but since it was still snowing and he only had one eight-inch gauge, that was not an option. He could only measure the melt water in the two-inch-diameter measuring tube to the nearest 0.10 inches, but that was better than nothing. What he found was between 0.60 and 0.70 inches of water from the melted snow core. That made more sense.

More than thirty minutes had elapsed since he started his observation, but finally he felt like he understood what was going on. Despite adverse weather and data that hadn't seemed right, he now had a handle on the storm. After he warmed up and dried off, he sat down with his observation form and filled in his daily report.

While no one will ever know for sure how much snow fell that day, the estimates made by the observer were systematic and logical. You can see that without thinking an observer could have seriously overestimated or underestimated the snowfall and most likely would have unknowingly undermeasured the precipitation. Over time, inaccuracies add up and result in very inferior climatic data.

The Day Snow Wouldn't Stick

It was nearly April and for the last week temperatures had soared into the seventies. The grass was turning green and the buds on many trees were about to burst open. But a strong Canadian cold front was bearing down from the north. After some morning sunshine and mild temperatures, a powerful thunderstorm rolled through at noon dropping lots of rain and even some small hail. Temperatures plunged. By mid-afternoon it was very cold and the sky looked Novemberish. At first, it looked like it might clear off, but then rolls of dark stratocumulus clouds began moving down from the north.

The thermometer read 40°F when the first snow squall hit. For thirty minutes it snowed hard and the north winds howled. The volunteer weather observer was working in the house, but she looked out to notice that snow was accumulating rapidly on the grass and on the north sides of trees and buildings – perhaps an inch. Other areas were only wet. Suddenly, the snow stopped as quickly as it began. An hour later the snow had almost entirely melted, just in time for the next squall. This time, the snow lasted close to an hour. The observer went outside with her ruler before the squall ended and noted nearly two inches of snow in a few protected areas on the greening yard. Then the sun came out briefly before setting in the west.

After supper, the observer stepped out to do her official weather observer duty. After recording daily high and low temperatures and resetting the thermometer, she went over to the precipitation gauge. She had forgotten to remove the funnel and inner tube, but no snow remained in the gauge. She used the calibrated measuring stick inserted into the narrow tube beneath the funnel and was amazed to find 0.78

inches of precipitation. Oh yes, most of that fell during the mid-day thunderstorm when the cold front arrived. She had hoped to use the water content to help make a reasonable estimate of the snowfall, but couldn't since that moisture was combined with the rain from the storm earlier in the day. She emptied the contents of the gauge and set it back in place.

Well, what about snowfall? It seemed like it snowed a lot, but as she looked on the ground around the weather station most of the snow already was gone except in a few protected spots. She had a snowboard that she had just painted the year before, but all the snow had melted from it. The observer went to one of the patches of remaining snow and was surprised by her measurement. The snow didn't seem very deep, but the ruler read 2.5 inches. She looked more closely. The snow was sitting on top of the fresh young grass while the ruler pushed right through the grass down to the ground. There was more than an inch of grass and air beneath the bottom of the snow. She jotted down 1.4 inches in her notepad and went inside.

As she sat down to record her daily observation and prepare to make a phone call to the National Weather Service, she remembered it might be helpful to melt a core sample of the remaining snow. She went back outside with a second eight-inch can and pushed it down on the snow. The sample stuck nicely inside the gauge, like cookie dough in a cookie cutter. She brought it inside and melted it. She was surprised to find 0.28 inches of water from that little layer of snow, but knew she had a good sample.

After some consideration, this is what the observer wrote down. For her daily precipitation total, she entered 0.78 inches although she knew she might have missed a little of the water from the snow since the funnel was in place and might have caused some of the snow to blow out of the gauge. For her daily snowfall total she wrote 2.6 inches, but with an asterisk. For her snow depth observation she recorded one inch. Over in remarks she wrote, "P.M. snow squalls, snow partially melted as it landed, snowfall estimated, maximum depth 2 inches, water content of remaining snow 0.28 inches."

Others might not have come up with the identical numbers, but the observer used good sense to make these estimates and noted them appropriately. If all observers follow these procedures, very consistent and representative results will be obtained.

These two stories really are not unusual. It could be and often is much worse. There could have been an uneven layer of old snow on the ground prior to these storms making it even more difficult to make a base measurement. Old snow also could have been blown and redistributed by the blizzard winds. There could have been more melting taking place causing greater differences between snowfall and snow depth. There also could have been rain before or during the blizzard to make it more difficult to use the precipitation data to recognize the low gauge catch.

In conclusion, remember these important points. Follow established procedures. Work hard to identify the best possible location for taking measurements. If possible, use a snowboard. Then, when the tough weather hits, use common sense and support your data with informative remarks. Maybe a blizzard will baffle you. Do your best. Sometimes, it is easier with blizzard snows to re-evaluate your measurement a few days after the storm. When you have seen the distribution of snow throughout your area – the size and extent of drifts and the depth of snow in protected areas, you may wish to improve your original estimate. That's OK. If you leave a trail of remarks, climatologists years later will be able to understand what you were trying to convey and will interpret the data appropriately. Also, whenever you have questions or are uncertain about your observations, please contact an expert.

Common Questions About Snow

Q: *Is it ever too cold to snow?*

A: No, it can snow even at incredibly cold temperatures as long as there is some source of moisture and some way to lift or cool the air. It is true, however, that most heavy snowfalls occur with relatively warm air temperatures near the ground – typically 15°F or warmer since air can hold more water vapor at warmer temperatures.

Q: *How big can snowflakes get?*

A: Snowflakes are agglomerates of many snow crystals. Most snowflakes are less than one-half inch across. Under certain conditions, usually requiring near-freezing temperatures, light winds, and unstable, convective atmospheric conditions, much larger and irregular flakes close to two inches across in the longest dimension can form. No routine measurement of snowflake dimensions are taken, so the exact answer is not known.

Q: *Why do weather forecasters seem to have so much trouble forecasting snow?*

A: Snow forecasts are better than they used to be and they continue to improve, but snow forecasting remains one of the more difficult challenges for meteorologists. One reason is that for many of the more intense snows, the heaviest snow amounts fall in surprisingly narrow bands that are on a smaller scale than observing networks and forecast zones. Also, extremely small temperature differences that define the boundary line between rain and snow make night-and-day differences in snow forecasts. This is part of the fun and frustration that makes snow forecasting so interesting.

Q: *For decades, I recall people telling me that there is one inch of water in every ten inches of snow that falls. Is that really true?*

A: The water content of snow is more variable than most people realize. While many snows that fall at temperatures close to 32°F and snows accompanied by strong winds do contain approximately one inch of water per ten inches of snowfall, the ratio is not generally accurate. Ten inches of fresh snow can contain as little as 0.10 inches of water up to 4 inches depending on crystal structure, wind speed, temperature, and other factors. The majority of U.S. snows fall with a water-to-snow ratio of between 0.04 and 0.10.

Q: *Does snow always get fluffier as temperatures get colder?*

A: No. Studies in the Rocky Mountains have shown that the fluffiest, lowest density (0.01 - 0.05) snows typically fall with light winds and temperatures near 15°F. At colder temperatures, the crystal structure and size change. At very cold temperatures (near and below 0°F) crystals tend to be smaller so that they pack more closely together as they accumulate producing snow that may have a density (water-to-snow ratio) of 0.10 or more.

Q: *Why is snow white?*

A: Visible sunlight is white. Most natural materials absorb some sunlight which gives them their color. Snow, however, reflects most of the sunlight. The complex structure of snow crystals results in countless tiny surfaces from which visible light is efficiently reflected. What little sunlight is absorbed by snow is absorbed uniformly over the wavelengths of visible light thus giving snow its white appearance.

Q: *Does snow change how sound waves travel? I recall waking up as a child and without even looking outside knowing that it had snowed just from the sounds I could hear.*

A: Yes, when the ground has a thick layer of fresh, fluffy snow, sound waves are readily absorbed at the surface of the snow. However, the snow surface can become smooth and hard as it ages or if there have been strong winds. Then the snow surface will actually help reflect sound waves. Sounds may seem clearer and travel farther under these circumstances.

Q: *Why is snow a good insulator?*

A: Fresh, undisturbed snow is composed of a high percentage of air trapped among the lattice structure of the accumulated snow crystals. Since the air can barely move, heat transfer is greatly reduced. Fresh, uncompacted snow typically is 90-95 percent trapped air.

Photo by Daniel B. Glanz

Bibliography

Many books, theses, journal articles, technical reports and manuscripts from scientific conferences were used in the preparation of this booklet. The following reference list includes only those few publications that were directly cited in this text. A more extensive bibliography was prepared as a part of this snow project and is available in print or electronically from the Colorado Climate Center.

References

1. Abbe, C., 1888: 1887 Annual Report of the Chief Signal Officer of the Army under the direction of Brigadier-General A.W. Greeley, Appendix 46. US Government Printing Office, Washington, DC, p. 385-386.

2. Goodison, B.E., 1978: Accuracy of Canadian snow gauge measurements. *J. Appl. Meteorol.*, **27**, p. 1542-1548.

3. Grant, L.O., and J.O. Rhea, 1974: Elevation and meteorological controls of the density of new snow. Adv. Concepts Tech. Study Snow Ice Resourc. Interdiscip. Symposium, National Academy of Science, Washington, DC, p. 169-181.

4. Hackett, S.W., and H.S. Santeford, 1980: Avalanche zoning in Alaska, USA. *J. Glaciol.*, **26**(94), p. 377-392.

5. Hagarty, J.H., 1963: History of Weather Bureau precipitation measurements. Key to Meteorological Records Documentation No. 3.082, US Department of Commerce Weather Bureau, US Government Printing Office, Washington, DC, 19 pp.

6. LaChapelle, E.R., 1962: The density distribution of new snow. USDA Forest Service, Wasatch National Forest, Alta Avalanche Study Center, Project F, Progress Report No. 2, Salt Lake City, UT, 13 pp.

7. Nakaya, U., 1954: *Snow Crystals: Natural and Artificial*. Harvard University Press, 510 pp.

8. Perla, R.I., and M. Martinelli, 1976: Avalanche handbook. Agricultural Handbook No. 489, USDA Forest Service, US Government Printing Office, Washington, DC, 238 pp.

9. Robinson, D. A., 1989: Evaluation of the collection, archiving, and publication of daily snow data in the United States. *Physical Geography*, **10**, p. 120-130.

10. US Department of Agriculture Forest Service, 1973: Instrumentation for snow, weather and avalanche observations. Snow Safety Guide No. 2. Rocky Mountain Forest and Range Experiment Station, Fort Collins, CO, 80 p.

11. US Department of Agriculture Forest Service, 1961: Snow avalanches. Agriculture Handbook No. 194, US Government Printing Office, Washington, DC, 84 pp.

12. US Department of Agriculture Natural Resources Conservation Service, 1988: Snow surveys and water supply forecasting. Agriculture Information Bulletin 536. NRCS Water and Climate Center, Portland, OR, 15 pp.

Other Suggested Reading

There is a great deal of wonderful literature available on the subject of snow. One can find detailed scientific reports, descriptive climatological analyses, historical accounts and personal reflections. Here are just a few titles you may wish to look up and read to pass the time while you wait for the next snow storm to arrive.

Colbeck, S.C., 1985: What becomes of a winter snowflake. *Weatherwise*, **38**, p. 312-315.

Finklin, A.I., and W.C. Fischer, 1990: Weather station handbook – an interagency guide to wildland managers. NFES #1140, National Wildfire Coordinating Group, Boise Interagency Fire Center, Boise, ID. (Excellent manual on instrumentation and weather observing procedures.)

Gray, D.M., and D.H. Male (editors), 1981: *Handbook of Snow*. Pergamon Press, Toronto, Canada, 776 pp. (an incredibly extensive technical reference).

Hall, D.K., and J. Martinec, 1985: *Remote Sensing of Ice and Snow*. Chapman and Hall, London, UK, 189 pp.

Judson, A., 1965: The weather and climate of a high mountain pass in the Colorado Rockies. Research Paper RM-16, USDA Forest Service, Fort Collins, CO, 28 pp.

LaChapelle, E.R., 1969: *Field Guide to Snow Crystals*. University of Washington Press, 101 pp.

Ludlum, D.M., 1966: *Early American Winters, Vol. I. 1604-1820*. American Meteorological Society, Boston, MA, 285 pp.

Ludlum, D.M., 1968: *Early American Winters, Vol. II. 1821-1870*. American Meteorological Society, Boston, MA, 257 pp.

McClung, D., and P. Schaerer, 1993: *The Avalanche Handbook*. The Mountaineers, Seattle, WA, 271 pp.

Mellor, M., 1964: Properties of snow. Cold Reg. Sci. Eng. Part III, Sec. A1, CRREL Monographs, US Army, Hanover, NH, 105 pp.

Schaefer, V.J., and J.A. Day, 1981: *A Field Guide to the Atmosphere*. The Peterson field guide series. Houghton Mifflin Company, Boston, MA, 359 pp.

US Dept. of Commerce, NOAA, National Weather Service Observing Systems Branch, 1989: Cooperative station observations. NWS Observing Handbook No. 2, Silver Spring, MD, 83 pp. (The instruction handbook for NWS cooperative observers.)

There are several periodicals that routinely contain information about snow:

Weatherwise (1948 - present) Heldref Publications, Washington, DC

Weather Watcher Review (1994 - present) International Weather Watchers, Springfield, MA

American Weather Observer (1984 - present) Belvidere, IL

Eastern Snow Conference (Annual Proceedings, 1944 - present)

Western Snow Conference (Annual Proceedings, 1933 - present)

Data Sources

Where to find snow data:

Local, regional, and national snow data, both historic and recent, are available through a number of sources. Data may be available in many forms ranging from original hand-written records or published data tabulations to computer files, CD-ROM products or on-line computer data sources. Many data sources can be accessed via the Internet. In addition to the list that follows, your local public or university library is also a good place to start.

National Climatic Data Center
151 Patton Avenue
Asheville, NC 28801-5001

National Snow And Ice Data Center
CB 449
University of Colorado
Boulder, CO 80309

National Weather Service, Office of Hydrology
National Operational Hydrologic Remote Sensing Center
1735 Lake Drive West
Chanhassen, MN 55317-8582

State Climatologists (list available from the National Climatic Data Center)

Regional Climate Centers: Northeast - Ithaca, NY, Southeast - Columbia, SC, Midwest - Champaign, IL, Southern - Baton Rouge, LA, High Plains - Lincoln, NE, Western - Reno, NV (list available from the National Climatic Data Center)

Natural Resources Conservation Service
Water and Climate Center
101 S.W. Main St., Suite 1600
Portland, OR 97204-3224

U.S. Army Cold Regions Research and Engineering Laboratory
72 Lyme Road
Hanover, NH 03755-1290

The National Climatic Data Center in Asheville, North Carolina.
Photo by Grant Goodge

Many organizations in the United States are involved in snow data collection. The following is not an exhaustive list.

U.S. Department of Commerce, National Weather Service and National Environmental Satellite, Data and Information Service

National Aeronautics and Space Administration

State water resources organizations (some states)

State departments of transportation

Universities and research centers

American Weather Observer Supplemental Observation Network

International Weather Watchers

Local weather spotters (often organized by media meteorologists or National Weather Service offices)

Ski areas

Snow Glossary

Ablation: The process of being removed. Snow ablation usually refers to removal by melting.

Accretion: Growth of precipitation particles by collision of ice crystals with supercooled liquid droplets which freeze on impact.

Blizzard: Winds of at least 35 miles per hour along with considerable falling and/or blowing snow reducing visibility to less than one-quarter mile for a period of at least three hours. Extremely cold temperatures often are associated with dangerous blizzard conditions, but are not a formal part of the modern definition.

Climatology: The science and study of climate – the aggregate of all weather conditions at a point over a long period of time.

Condensation: The process in which water vapor becomes liquid.

Dendrite: Hexagonal ice crystals with complex and often fernlike branches.

Depth hoar: Large (one to several millimeters in diameter), cohesionless, coarse, faceted snow crystals which result from the presence of strong temperature gradients within the snowpack.

Evaporation (water): The physical process in which liquid water changes into a gas.

Graupel: Snowflakes that become rounded pellets due to riming. Typical sizes are two to five millimeters in diameter (0.1 to 0.2 inches). Graupel is sometimes mistaken for hail.

Metamorphism: Changes in the structure and texture of snow grains which results from variations in temperature, migration of liquid water and water vapor, and pressure within the snow cover.

Polycrystal: A snowflake composed of many individual ice crystals.

Precipitation: The accumulated depth of rain or drizzle and also the melted water content of snow or other forms of frozen precipitation, including hail.

Rime: A deposit of ice formed when supercooled water droplets freeze on contact with an object.

Saturation Vapor Pressure (water): The maximum amount of water vapor necessary to keep moist air in equilibrium with a surface of pure water. This is the maximum water vapor the air can hold for any given combination of temperature and pressure.

Snowboard: A solid, flat, white material, such as painted plywood, approximately two feet on each side, that is laid on the ground or on the surface of the snow by weather observers to obtain more accurate measurements of snowfall and water content.

Snowbursts: Very intense showers of snow, often of short duration, that greatly restrict visibility and produce periods of rapid snow accumulation.

Snow core: A sample of snow, either just the freshly fallen snow or the combined old and new snow on the ground, obtained by pushing a cylinder down through the snow layer and extracting it.

Snow density: The mass of snow per unit volume which is equal to the water content of snow divided by its depth.

Snow depth: The combined total depth of both old and new snow on the ground.

Snowfall: The depth of new snow that has accumulated since the previous day or since the previous observation.

Snowflake: A cluster of ice crystals that falls from a cloud.

Snow flurries: Snow that falls for short durations and which often changes in intensity. Flurries usually produce little accumulation.

Snow load: The downward force on an object or structure caused by the weight of accumulated snow.

Snow water equivalent: The water content obtained from melting accumulated snow.

Snowpack: The total snow and ice on the ground, including both new snow and the previous snow and ice which has not melted.

Snow squall: A brief, but intense fall of snow that greatly reduces visibility and which is often accompanied by strong winds.

Sublimation: The process in which ice changes directly to water vapor without melting, but also in meteorology the opposite process in which water vapor is transformed to ice (also called deposition).

Supercooled: The condition when a liquid remains in the liquid state even though its temperature is below its freezing point.

Supersaturation: The condition which occurs in the atmosphere when the relative humidity is greater than 100 percent.

Surface Hoar: The deposition (sublimation) of ice crystals on a surface which occurs when the temperature of the surface is colder than the air above and colder than the frost point of that air.

Vapor Pressure: The pressure exerted by water vapor molecules in a given volume of air.

For further definitions, refer to the Glossary of Meteorology available from the American Meteorological Society, Boston, Massachusetts.

Credits and Acknowledgements

Many snow experts and snow lovers have shared some of their knowledge or resources to help with this snow booklet. So many people have helped in some way that this has truly been a team effort. To all of you – many, many thanks for your help.

Funding: This project was made possible by funding provided by the U.S. Department of Interior, Bureau of Reclamation Flood Hydrology Group under the supervision of Louis Schreiner. A large amount of special project assistance was also provided by the National Oceanic and Atmospheric Administration National Climatic Data Center and the Colorado State University Agricultural Experiment Station.

Data providers: The USDA Natural Resources Conservation Service, Greg Spoden, Dan Leathers, David Robinson, Knox Williams, Roy Bates, L. David Minsk, Dwight Pollard, Richard Armstrong, Grant Goodge. A special thanks to Richard Heim of the National Climatic Data Center who spent many hours assembling the 1961-1990 snowfall statistics for the United States.

Bibliography: Richard Armstrong, Tom Schwein, Lee Larson, Andy Standley, Phil Farnes, Tom Schmidlin, and many others who brought reference material about snow to our attention. Also Odie Bliss and Jim Harrington for assembling and formatting the extended bibliography that supported this project.

Manuscript review: Tom McKee, Ned Guttman, John Hughes, Grant Goodge, Richard Heim, Ray Kowrach, Tom Blackburn, Lee Larson, National Weather Service Headquarters, Richard Armstrong, Lou Schreiner, Jack McPartland, Sam Colbeck, James Wirshborn, Jim Harrington, Tom Carroll, Owen Rhea, Pam Knox, David Robinson, Phil Pasteris, and Dan Leathers.

Photographs: Grant Goodge, Walter Johnson, Ken Dewey, Dan Glanz, Richard Keen, Tom Carroll, Richard Armstrong, Jim Harrington, Robert Muller, Nebraska Historical Society, Colorado Historical Society, Garry Schaeffer, James Wiesmueller, Allen Dutcher, Wayne Wendland, Tom Blackburn, Tom Dietrich, and Charles Kuster.

Encouragement: Larry Mooney, Hal Klieforth, Kelly Redmond, Phil Pasteris, Tom Blackburn, Ken Hadeen, John Hughes, Tom McKee, American Association of State Climatologists.

For all the work it took to put the final publication together, we would like to thank Jim Harrington for countless hours tracking down and analyzing data, Odie Bliss for reformatting the manuscript time after time, Natalie Marquez for helpful assistance, John Kleist for computer support, Judy Sorbie-Dunn for her art, drafting and layout work, Lisa Helme for editing the text, and Kathy Hayes and Jeannine Kline at Colorado State University Publications and Printing for getting the job done.

Special personal thanks to Grant Goodge of the National Climatic Data Center and David Robinson, New Jersey State Climatologist, Rutgers University for breaking from their busy schedules time after time to offer assistance, encouragement and wisdom about snow. And finally, thanks to our wives and families who put up with us during this stage of our lives.

Snow Memories

There is something about snow that enhances our ability to remember. Our life experiences, both sad and joyous, are more deeply etched in our minds if accompanied by snow. Ask any person you meet to recall some past experience that involved snow, and they, most likely, will begin sharing a story complete with intricate detail, intense feelings, and perhaps a touch of exaggeration. It will be a story worth listening to. And if you are prone to poetry, snow will make you want to write something, even if you are the only one who ever reads it.

There are thousands of stories, pages of poetry, and countless cartoons that somehow involve snow, but there is nothing richer than a cherished recollection of a snowstorm from your youth. If somehow we could each put our fondest or most poignant snow-related memory onto paper, what a fine gift that would be to pass on to our children and grandchildren.

The following recollection is one of several sent to the Climate Center. Perhaps it will spur your memory.

The Armistice Day Blizzard of 1940: Wesley Gulliksen Remembers

It was an unseasonably warm 66 degrees that afternoon in Northfield, Minnesota – November 11, 1940. At the time I was a student at St. Olaf College, and I remember the day being so pleasant that my housemates and I went to a nearby field in our gym shorts to play touch football.

Our game was cut short when there was a sudden shift in the wind to the northwest and a noticeable chill in the air. In the distance an ominous cloud line was getting larger by the minute, a type of cloud formation that northerners refer to as a "snow bank."

Within an hour the temperature dropped to freezing. The wind became a gale, and the snow began. It was the beginning of what was later called the Armistice Day Blizzard.

The next morning the snow was three feet deep on the level. On the windward side of our rooming house, drifts were up to our second story windows. The temperature was 42 degrees below zero – a drop of 108 degrees.* Over the radio we learned the winds ranged from 50 to 80 miles per hour. Wind chill was not reported in those days; however, based on temperature and wind speed, the wind chill may have been as low as -120°F.

The blizzard centered in our northern states, from the Dakotas to Ohio. Livestock perished by the hundreds. Worthington, Minnesota, which claimed to be the "Turkey Capital of the World" had their entire Thanksgiving harvest of 500,000 birds wiped out.

A 10-foot subalpine fir tree is almost buried following a record-breaking 112 inches of snowfall at Steamboat Springs, Colorado, during January 1996.

Photo by Arthur Judson

* Based on Minnesota official climatological records, this well-known storm began with temperatures in the mid 40s and dropped to +1° Fahrenheit.

Many motorists died on the highways. Stalled by drifts and zero visibility, they tried to survive by keeping car motors and heaters running. When the snow covered the cars' tailpipes, they died from carbon monoxide. On the Great Lakes a half dozen freighters were beached or sunk.

I had two uncles who skippered such freighters on the Lakes. One of them, Captain Carl Christensen, was bringing his ship to its winter port in Milwaukee. He entered Lake Michigan through the Straits of Mackinac when the storm hit. From that point his ship was carried by the storm as if it had no rudder. He finally got his vessel under control a dozen hours and 323 nautical miles later, just ten miles out of Chicago. He later recalled that in all his years, both on the High Seas and the Great Lakes, that was the worst storm he ever saw.

One of my cousins, Norman Gulliksen, was on a duck hunt with some friends near De Soto, Wisconsin, on the banks of the Mississippi River. He gives the following account:

On November 11 the temperature was 73° at 8:00 AM, dropping to freezing by noon. We got off the river because some old-time hunters drove by our blind yelling as they passed us, 'There's a bad storm coming in through the Dakotas.'

By 11:00 AM the western sky was black, and the wind was at least 40 miles per hour. We hurriedly began to gather up our decoys. As we pulled them in, the anchor cords froze solid before we could wind them up. Far out on the river we saw two hunters tip over in their skiff. I'm sure they never made it. Later we learned that approximately 30 hunters froze to death in that area.

We drove out of there as fast as we could, and we managed to get about 140 miles, to Cottage Grove just east of Madison, before our car quit. During a ten minute stop in a cafe, the oil in the engine turned to tar. It was before the days of all-weather oil. So we waited out the storm in an old motel.

We finally reached Milwaukee on the 13th. The temperature there had been down to -48°, but by the 13th it had crept up to -30°. Then more snow and wind came. There were nine to 16 foot drifts. Work places and transportation in Milwaukee were shut down for four days. The winds blew out most of the store windows on downtown Wisconsin Avenue.

Meanwhile, back in Northfield, Minnesota, we were snowbound for six days. On the fourth day we ran out of food, so we risked a block walk to a small grocery where, luckily, the owner lived in the rear of his store building. Our winter dress covered every inch of our bodies, with only a slit opening at eye level for visibility. Even so, the icy wind seemed to suck the very air from our lungs.

On the seventh day the gale ended. Winds were calm. There wasn't a cloud in the sky, and the sun shone brightly. It was still 18 degrees below zero, but we walked downtown and back to get rid of our "cabin fever."